GIS
TUTORIAL
for
ArcGIS® Pro
2.8

GIS
TUTORIAL
for
ArcGIS® Pro
2.8

WILPEN L. GORR

KRISTEN S. KURLAND

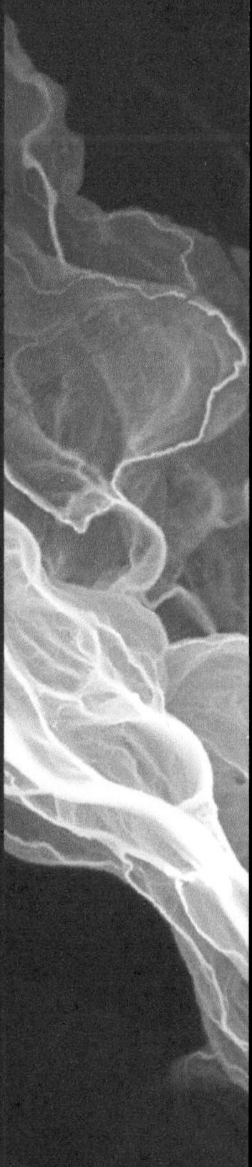

Esri Press
REDLANDS | CALIFORNIA

The cover design is based on the map
The Bare Earth by the Washington Geological Survey,
at esri.com/en-us/maps-we-love/gallery/.

Esri Press, 380 New York Street, Redlands, California 92373-8100
Copyright © 2021 Esri
All rights reserved.

Printed in the United States of America
25 24 23   5 6 7 8 9 10

ISBN: 9781589486805
Library of Congress Control Number: 2021939321

Contents

Preface

Welcome to GIS Tutorial for ArcGIS® Pro 2.8. This fourth-edition step-by-step workbook focuses on ArcGIS Pro but also covers ArcGIS® Online and some of its major apps for mobile computing, including ArcGIS® StoryMaps™, ArcGIS® Dashboards, and ArcGIS® Collector.

ArcGIS Pro is Esri's powerful single desktop GIS application, with an intuitive user interface for creating, using, and analyzing professional 2D and 3D maps. ArcGIS Pro's tasks, ModelBuilder™ models, and apps also allow you to build operational systems that help organizations use GIS daily. You create tasks to guide complex workflows and build models to automate multiple geoprocessing steps, all without computer coding. The apps provide GIS results and capabilities to all who need them.

ArcGIS Online is software as a service (SaaS) for making, using, analyzing, and sharing maps. Using this service, you can connect people, locations, and data using interactive maps. You can easily publish maps authored in ArcGIS Pro to ArcGIS Online and integrate them with maps from ArcGIS® Living Atlas of the World, available in ArcGIS Online. Living Atlas, curated by Esri, is the world's foremost collection of ready-to-use maps. With your finished maps stored in ArcGIS Online, you can easily share them in ArcGIS Online apps.

The StoryMaps app allows you to incorporate text, charts, and other content with interactive maps to tell a story or make a report on a GIS project. Dashboards monitors an organization's demands for services and performance with maps, statistics, charts, lists, and queries. Collector allows for collecting and updating spatial and attribute data in the field using mobile computing devices.

This book is a complete learning system for GIS, including features we have updated in more than 25 years of teaching GIS using Esri products and writing tutorials. Our books are used successfully by students in classes as well as by self-learners. We've taught high school students, in-career professionals, undergraduate students, master's students, PhD students, and distance-learning students across many disciplines. We teach using a combination of lectures and lab sessions, and we always include student-designed GIS projects as a final requirement in our courses. Our interactions with students are important sources of ideas and feedback for our books.

This book includes wide-ranging real-world data and GIS applications that require integration of different GIS processing steps and workflows to solve realistic problems. This book addresses many topical issues, including the following examples:

- Accessing urgent-care health clinics by low-income populations
- Finding employment prospects in the arts fields across US cities, including the size of arts communities, average incomes, and cost of living
- 3D modeling of city buildings using lidar data for city planning
- Building a system for scheduling and routing graffiti removal in a city
- Location analyses
  - Placing defibrillators outdoors to resuscitate heart attack victims in a city
  - Locating farmers' markets as part of a city's solution to its "food desert" problem
- Digitizing a university campus's buildings and walkways from satellite images for use in master planning

- Data mining crime patterns in a city for insights into criminal behavior

The book presents important concepts at the start of each chapter or at the start of tutorials. Sections on concepts are separated from tutorial steps to ensure continuous, hands-on computer work.

Short chunks of step-by-step tutorials are illustrated with graphics and brief comments when helpful. We try to limit these short sections to a maximum of 10 steps each to help you stay focused. These small chunks facilitate learning and retention and provide a sense of accomplishment.

In Your Turn assignments within tutorials, you will repeat steps just taken but in slightly modified ways. These short assignments help you begin to internalize steps and workflows.

In ArcGIS Online (go.esri.com/GISTforPro2.8Data), you will find related assignments and data to help you apply each chapter's concepts and workflows independently. The assignments and data are in the Learn ArcGIS organization's GIS Tutorial for ArcGIS Pro 2.8 (Esri Press) group. The tutorial steps expose you to GIS and serve as a reference, whereas the assignments provide additional resources for more in-depth learning. In the classroom, instructors likely will ask you to turn in assignments for grading. Resources (go.esri.com/GISTforPro2.8) include links for video lectures and lecture slides covering the entire book, especially useful for self-learners. This book also comes with a free software trial. Use the EVA code inside the back cover of this book, and go to go.esri.com/EVAcode.

The four parts of this book are organized in a sequence that best motivates and facilitates your learning of GIS.

In part 1, "Using, making, and sharing maps," the first three chapters of the book introduce you to the ArcGIS user interface through finished maps. You will symbolize a range of map types using cartographic principles and build map layouts and use StoryMaps to present your results.

The five chapters in part 2, "Working with spatial data," prepare you for finding and understanding spatial data. You will store and process spatial data in file geodatabases, use spatial data in geoprocessing for mapping and analysis, digitize your own spatial data, and geocode tabular data for mapping.

The three chapters in part 3, "Applying advanced GIS technologies," focus on analyzing spatial data and maps to solve problems. You will apply several unique GIS methods for analyzing spatial relationships, including buffers, service areas, facility location, and data clustering. You will use raster GIS (which works with satellite images and other continuous data) to analyze demand for services. Part 3 concludes with an exploration of 3D GIS to model urban redevelopment projects.

In part 4, "Managing operational systems with GIS," the final two chapters of the book provide hands-on experience in building operations management systems. In chapter 12, you create tasks and a ModelBuilder model to prepare and publish weekly graffiti location data on a map. You also create an operation view using Dashboards with the published data for police to use in preventing graffiti by serial artists. In chapter 13, you build three ModelBuilder models for a public works supervisor to use in scheduling and routing graffiti removal. You will prepare a map in ArcGIS Pro and ArcGIS Online for use in the Collector app for the supervisor to enter assessments from in the field on a mobile device for graffiti removal.

GIS is our favorite subject to teach and a favorite class for our students to take. If you have any questions or feedback, you can reach us via email at gorr@cmu.edu or kurland@cmu.edu. Bon voyage!

Acknowledgments

We would like to thank all who made this book possible.

We have taught GIS courses at Carnegie Mellon University since the late 1980s, always using our own lab materials. With the feedback and encouragement of students, teaching assistants, and colleagues, we eventually wrote what became the GIS Tutorial series of workbooks, leading to this book. We are forever grateful for that support.

Faculty members of other universities who have taught GIS using our books have also provided valuable suggestions and feedback. They include Luke Ward of Rocky Mountain College; Irene Rubinstein of Seneca College; An Lewis of the University of Pittsburgh; George Tita of the University of California, Irvine; Walter Witschey of Longwood University; Jerry Bartz of Brookhaven College; and James Querry of Philadelphia University.

We are grateful to the many public servants and vendors who have generously supplied us with interesting GIS applications and data, including Eli Thomas of the Allegheny County Division of Computer Services; Kevin Ford of Campus Design and Facility Development, Carnegie Mellon University; Barb Kviz of the Green Practices program, Carnegie Mellon University; physicians at Children's Hospital of Pittsburgh of UPMC; Erol Yildirim, Council for Community and Economic Research; Mike Homa of the Department of City Planning, City of Pittsburgh; staff members of the New York City Department of City Planning; Wendy Urbanic, Pittsburgh 311 Response Center; Bob Gradeck, Western Pennsylvania Regional Data Center; Michael Radley of the Pittsburgh Citiparks Department; Pat Clark and Traci Jackson of Jackson Clark Partners; Maurie Kelly of Pennsylvania Spatial Data Access (PASDA); staff of the Pennsylvania Resources Council; Kirk Brethauer of the Southwestern Pennsylvania Commission; Steve Benner of Pictometry International Corp.; and employees of several spatial data vendors, including Esri, HERE Technologies, the National Geospatial-Intelligence Agency, US Geological Survey, and the National Park Service.

Many technical and expert GIS staff members of Esri reviewed the first draft of this book, and we are grateful for their comments, corrections, and clarifications. It was a pleasure working with these dedicated and talented professionals. Any remaining errors are ours. Finally, we are much indebted to the wonderful staff at Esri Press for their editorial expertise, beautiful design work, efficient production, and distribution of our book.

PART 1

Using, making, and sharing maps

Essentials for getting started

About this edition

This edition of *GIS Tutorial for ArcGIS® Pro 2.8* has been tested for compatibility with ArcGIS Pro 2.8. For this edition, new graphics have been created to reflect the latest software. Some steps have changed to provide clarity and reflect changes in how the software works.

Hardware and software requirements

To perform the exercises in this book, you will need ArcGIS Pro installed on a computer that is running the Windows operating system, an internet connection, a mobile device, and a web browser to access ArcGIS Online or other software as a service or apps. Earlier software versions may not be fully compatible with exercise data and may not operate as described in the exercises. Hardware requirements are available at go.esri.com/ArcGISProSysReqs.

Licensing the software

Use an existing license

If you have existing credentials (or can obtain credentials from your educational institution or other organization) that provide access to the required elements of ArcGIS, you may use those credentials and proceed.

Use an evaluation (EVA) code

This print or e-book comes with an EVA code that will grant you a fully functional, nonrenewable, 180-day license. You can find the code inside the back cover of the print book. You can access the e-book code solely through the e-book delivery platform, Vital Source, and view it after renting or purchase. Activate your code and license your software at go.esri.com/EVAcode.

Use a trial version

Additional trial account options can be found at arcgis.com.

Installing the exercise data

The exercise data for this book is available at go.esri.com/GISTforPro2.8Data. It is shared with the ArcGIS Online group GIS Tutorial for ArcGIS Pro 2.8 (Esri Press) in the Learn ArcGIS organization. Download the tutorial data and store it on your computer.

How to use this book and access additional resources

This book is designed for chronological progression—earlier chapters have more explicit instruction than later chapters. Also, exercises within chapters typically build upon each other, so it is advisable to perform all the exercises within a chapter in numerical order. The ArcGIS Pro Help documentation provides comprehensive descriptions of software concepts and tools at go.esri.com/help.

Esri Press website
Instructor and student resources are available from the book's web page at go.esri.com/GISTforPro2.8.

Feedback and updates
Feedback, updates, and other useful information are available at Esri Community, the global community of Esri users for finding solutions, sharing ideas, and collaborating to solve problems at go.esri.com/EsriBooks.

Data license agreement
Downloadable data that accompanies this book is covered by a license agreement that stipulates the terms of use.

Introducing ArcGIS

LEARNING GOALS

- Get an introduction to ArcGIS.
- Get an introduction to the ArcGIS Pro user interface.
- Learn to navigate maps.
- Work with tables of attribute data.
- Get an introduction to symbolizing and labeling maps.
- Work with side-by-side 2D and 3D maps.
- Publish a map to ArcGIS Online.
- Configure maps in ArcGIS Online.
- Use ArcGIS Explorer on a mobile device.

Introduction

ArcGIS is an integrated collection of GIS software packages and apps developed by Esri that work seamlessly across desktop computers, the internet, and mobile devices. The tutorials in this first chapter will familiarize you with some major components of this software: ArcGIS Pro, ArcGIS Online, and ArcGIS Explorer. You'll use additional ArcGIS apps and packages in other chapters.

ArcGIS Pro, the major software taught in this book, is a 64-bit desktop GIS application that uses a ribbon interface for 2D and 3D map authoring, analysis, and web publishing. The interface makes relevant tools visible and available for whatever work you're doing in GIS. ArcGIS Online is Esri's cloud solution for interactive web mapping and spatial data sharing. Maps that you create in ArcGIS Pro can be published to ArcGIS Online. Then, once in ArcGIS Online, maps can be accessed in web browsers and in mobile-device apps. ArcGIS Explorer is a simple interactive viewer for your online maps.

In this chapter, you will work with a finished map that has the locations of urgent health care clinics in Allegheny County, Pennsylvania. These clinics are federally qualified health centers (FQHCs) that provide subsidized health care for underserved populations and MedExpress clinics that provide private health care. In part, both FQHCs and MedExpress centers are low-cost alternatives to hospital emergency rooms. You will examine the finished map's components while navigating through user interfaces and around mapped features. In the process, you'll learn that both the publicly funded and private-sector urgent health care clinics are well located in interesting spatial patterns.

Tutorial 1-1: Overview of ArcGIS Pro

Before starting work on your computer, review key terminology for ArcGIS Pro projects and spatial data.

- A *project* is a file, with the extension.aprx, that contains one or more maps and related items. For example, you'll open project Tutorial1-1.aprx in ArcGIS Pro after this introduction. The project has two maps, Health Care Clinics and Health Care Clinics_3D, plus other project items.

- A project has a home folder of your choice. The home folder of Tutorial1-1.aprx is Chapter1\ Tutorials. If you installed this book's data on the C drive of your computer, the location of the Tutorial1-1.aprx project is C:\EsriPress\GISTforPro\Chapter1\Tutorials\Tutorial1-1.aprx.

- A *file geodatabase* is a folder, with the extension.gdb, that stores one or more feature classes, rasters, and other related files. Although there are many other file formats for storing spatial data, the file geodatabase is a preferred Esri format. The data used in Tutorial1-1.aprx is in the file geodatabase Chapter1.gdb, stored in the Chapter1\Tutorials folder on your computer. A project does not store spatial data used to make maps but instead stores connections to spatial data, such as a file geodatabase, that is stored elsewhere on your computer, ArcGIS Online, or other locations.

- A *feature class* is composed of spatial data and is the basic building block of GIS for storing features that can be graphically displayed on a map. Feature classes have corresponding attribute data for each feature. For example, Chapter1.gdb has a feature class named FQHC that has point locations for all FQHCs in Allegheny County, along with attribute data including the FQHC name and address. Chapter1.gdb has many more feature classes, one of which is Municipality, which has boundaries for all municipalities in the county (including the city of Pittsburgh). Yet another feature class is Streets, which has centerlines for all streets in the county.

- A *raster dataset* (or raster) is the other major type of spatial data for mapping. Quite often, a raster is a stored image made up of pixels—square areas with assigned colors so small that you can't see them individually until you zoom in close. In general terms, a raster is a rectangular table with numbers in cells (the pixels), with cells referenced to geographic coordinates. For images, the stored numbers correspond to assigned colors.

- A *map layer* is a feature class or raster as visualized in a map, and a map is a composition of map layers overlaying each other. You choose and symbolize the layers to serve a given purpose.

It's important to understand that an ArcGIS Pro project is a file that stores your maps, but the spatial data (feature classes and rasters) that maps contain is stored elsewhere on your computer, a local area network, or in the cloud on the internet.

Open the Tutorial 1-1 project

This book's tutorials have prebuilt projects that you open and use or modify to complete lessons. So you'll start by opening ArcGIS Pro and opening a project.

1. Browse to go.esri.com/GISTforPro2.8Data to download the tutorial data for the book, which is hosted in the group GIS Tutorial for ArcGIS Pro 2.8 (Esri Press). Download and extract the files to C:\.

2. Start ArcGIS Pro on your computer.

3. Sign in with your ArcGIS account user name and password.

4. Click Open Another Project, browse on your computer to C:\EsriPress\GISTforPro\Chapter1\ Tutorials, and double-click Tutorial1-1.aprx. If you don't see the Open Another Project link, you can widen the ArcGIS Pro window. The project opens and displays a map, Health Care Clinics, which includes 14 symbolized map layers. You can turn layers on and off by selecting and clearing the check boxes next to their names. Only one layer is selected, Population Density. You'll select more momentarily. If you do not see the Contents pane, you'll open that pane in the next step.

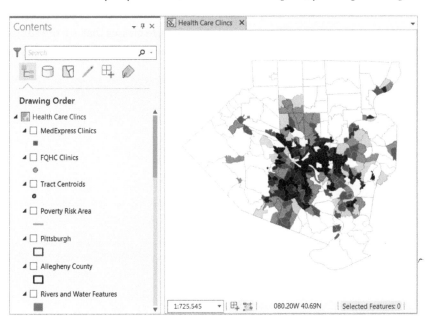

5. If the Contents pane is open, click the Close button in the upper-right corner of the pane.

6. On the View tab, go to the Windows group, and click Contents to open the Contents pane. The Contents pane normally docks on the left. Optionally, you can float any pane (you will open several other panes later in this chapter) by right-clicking the top of the pane and clicking Float or by clicking and dragging the top of the pane outward. You can experiment by right-clicking the top of the pane and clicking Dock to redock the Contents pane on the left.

7. **On the Map tab, in the Navigate group, click the Full Extent button** ⊕. Clicking the Full Extent button zooms the map to the full extent of the data. If the map was zoomed in to a small area, clicking the Full Extent button would display the entire map.

8. **On the Project tab, click Save As, browse to Chapter1\Tutorials, and save as** Tutorial1-1YourName .aprx **(substitute your name for YourName)**. You'll generally save provided projects this way, so that if you make a major mistake, you can start again with the original project.

Add and remove a basemap

A basemap is a layer that helps orient the map user to the location. Map designers place additional feature classes on top of a basemap to provide specific information for visualization, analysis, or solving a problem. Although you can create your own basemap, Esri provides the basemaps that you'll use in this book from the Esri web portal, ArcGIS Online. By default, projects created using ArcGIS Pro have the topographic basemap added to the bottom of the Contents pane. Additional basemaps are available.

1. **On the Map tab, in the Layer group, click Basemap.** You will see a variety of basemaps—available basemaps will depend on licensing. Many of the maps you'll build in this book will use the Light Gray Canvas basemap because you'll reserve color for feature classes that are the subject of the map. Basemaps in the background will provide spatial context for the locations of subject features.

2. **Click the Streets basemap to add the basemap to your map.** The Population Density map covers most of the Streets basemap. Because the areas of the basemap that are visible outside of Allegheny County do not match Population Density, or otherwise add useful information, you will remove the basemap for now.

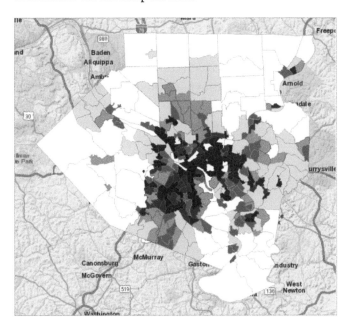

3. **If necessary, scroll to the bottom of the Contents pane, right-click World Street Map, and click Remove.**

YOUR TURN

The Your Turn assignments in this book will ask you to repeat the steps just completed, but with some modifications. These assignments will help you retain the workflows in the steps. Often, you will need to complete the Your Turn assignments so you can use their results in the next tutorial steps, so do not skip Your Turn assignments.

For this Your Turn assignment, add and remove several basemaps of your choice. You will notice that some basemaps, such as the light- and dark-gray canvases, add a labeling layer at the top of the Contents pane. When you remove the last basemap, remove the labeling layer, if necessary.

Turn layers on and off

The order of drawing by ArcGIS Pro is from the bottom up in the Contents pane. So feature classes that cover areas, such as Population Density, must go on the bottom, and other feature classes that could be covered up, such as FQHC Clinic points, must go higher up and on top of other feature classes.

1. **In the Contents pane, scroll down to see the legend for Population Density.** The check mark on the left of Population Density indicates that the feature class is turned on. This feature class represents persons per square mile in 2010 by census tracts in numeric classes, with uniform widths of 1,000 people per square mile. Census tracts are statistical areas intended to represent neighborhoods with about 4,000 people, although population tracts can vary widely in population.

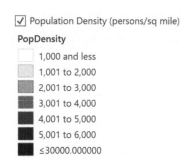

2. **Continuing in the Contents pane, select the small boxes on the left of MedExpress Clinics, FQHC Clinics, and Poverty Risk Area.** The three feature classes you just turned on are the subject of this map and show the locations of urgent health care clinics relative to poor areas. Right away, you can see that the subsidized FQHCs are concentrated in high-population density (urban) and poverty risk areas, whereas the private-sector MedExpress clinics are mostly spread out in low-population density areas (suburbs). Areas inside the Poverty Risk Area polygons have high proportions of poor populations.

3. **Turn on feature classes that provide the spatial context of where subject features are located: Allegheny County, Pittsburgh, Rivers and Water Features, and Streets.** Streets, an important spatial context feature class, will not display until the map is zoomed in to a small area (you'll learn about zooming later in this chapter). There are too many detailed streets for viewing at full extent. Next, to make the point that ArcGIS Pro draws from the bottom up in the Contents pane, you'll temporarily drag Population Density up the pane to cover other feature classes.

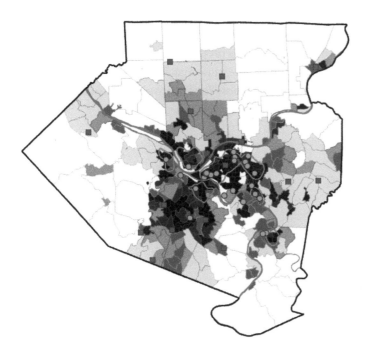

Move feature classes in the Contents pane

1. **Drag Population Density to the top of the Contents pane under Health Care Clinics.** Now this feature class covers all other feature classes in the map.

2. **Drag Population Density back to just above the Poverty Index feature class.**

3. **Click the Save button** 🖫 . The map, composed of feature classes currently displayed, has useful information that you can share by publishing the map to ArcGIS Online. From there, the map could be viewed on a mobile device using ArcGIS Explorer. You'll publish a map similar to this one in tutorial 1-4 and use the published map with Explorer in tutorial 1-5.

Examine the Catalog pane, and open and export a map layout

The Catalog pane provides access to all components of an ArcGIS Pro project.

1. **If the Catalog pane is not already open, click the View tab, and click the Catalog pane.** The Catalog pane appears. If the pane is not docked on the right in the ArcGIS Pro window, right-click it at the top and click Dock.

2.   **In the Catalog pane, click the arrows on the left of both the Maps and Layouts folders to expand the folders—revealing what's been built so far for this project.** You are viewing the Health Care Clinics map, but you will also view a 3D version of the same map later in this chapter. Next, you will open the layout for FQHC and MedExpress Clinics. You'll learn about the other project components (toolboxes, databases, and so on) later in this book as needed.

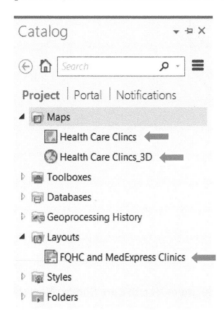

3.   **In the Catalog pane under Layouts, double-click FQHC and MedExpress Clinics.** ArcGIS Pro displays the layout on a new tab. The map is the main element of a layout, which also includes map surrounds such as the title, legend, and graphic map scale. These elements make the map suitable for use as a figure in a report or on a slide in a presentation. You'll learn how to create layouts from maps in chapter 3.

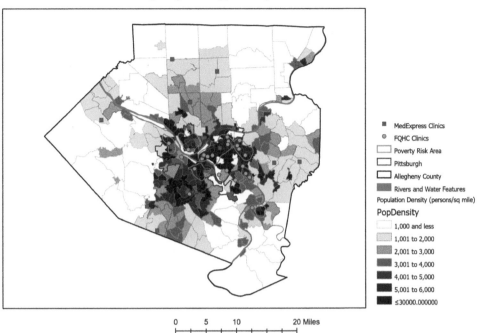

**Poverty Areas and Population Density by Census Tract
in Allegheny County, Pennsylvania**

4. **At the upper right of the Catalog pane, click the Auto Hide button 📌 to temporarily hide the pane.** You can restore and hide the Catalog pane as needed by clicking the Catalog button.

5. **On the Layout tab, in the Navigate group, click the Full Extent button 🔳.** Next, you'll export the map as an image file.

6. **On the Share tab, in the Output group, click Export Layout, and change File Type to JPEG.**

7. **Change the file name to** HealthClinics **and save to Chapter1\Tutorials. Set** 300 DPI **(dots per inch) for Resolution, and click Export.** Open the output image file in a photo viewer (double-click HealthClinics.jpg). At 300 DPI, this is a high-quality image that you could print or insert into a Microsoft Word document or Microsoft PowerPoint presentation. For use in a website or mobile device, the recommended resolution is much lower (72 DPI) to keep file sizes small and loading times fast.

8. **Close the Export Layout pane and close the FQHC and MedExpress Clinics tab to close the layout, and click the Save button to save your project.**

YOUR TURN

Turn on the FQHC Buffer and MedExpress Buffer feature classes. The buffers are one-mile radius circles constructed with health care clinics at their centers. The rationale for choosing a one-mile buffer radius is that this radius is commonly used to determine accessibility to grocery stores in urban areas. You can assume that what works for grocery store accessibility also works for health care facilities. Notice that the buffers are partly transparent so that you can see the population density below them. Next, open the layout. ArcGIS Pro automatically adds the newly displayed feature classes to the layout's map and legend. As stated earlier, the FQHCs appear well located within poverty and densely populated areas, whereas the MedExpress facilities are mostly scattered in suburbs surrounding Pittsburgh. Why do you think these patterns exist? When finished, close the layout and save your project.

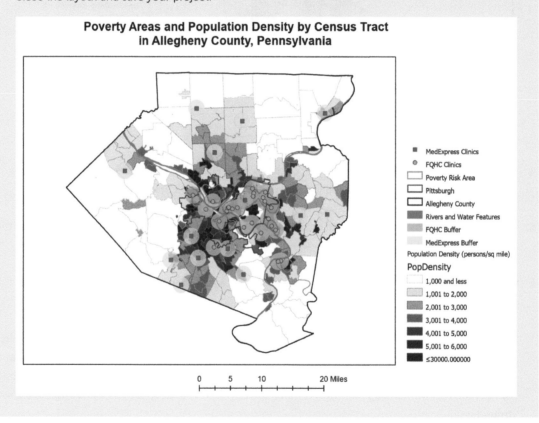

Poverty Areas and Population Density by Census Tract in Allegheny County, Pennsylvania

Tutorial 1-2: Navigate ArcGIS Pro

Map reading in GIS depends on varying location and scale (where and how far you are zoomed in) and using the attribute data of spatial features. You can zoom in to any part of a map, drag (pan) the map to a different location, and zoom back out. You can set some feature classes to display only when they are zoomed in to a certain scale and beyond, such as streets, and you can clear the display for other feature classes. You can go to preset locations and scales using spatial bookmarks. You can read the attribute data of any feature by clicking the feature to get a pop-up window. Last, you can search for features by using attribute values such as the name of a street.

Open the Tutorial 1-2 project

1. Click the Project tab.

2. Click Open and browse to Chapter1\Tutorials.

3. Open Tutorial1-2.aprx, and save the tutorial as Tutorial1-2YourName.aprx in **Chapter1\Tutorials.**

4. Click Full Extent to zoom in on the map.

Use a pop-up window

1. Click the Map tab on the ribbon, and click the Explore button.

2. On the map, click the MedExpress Clinic farthest to the left (west) to see a pop-up window with attribute data for that feature. Click the pop-up's website hyperlink, and when you finish, close your browser.

Private-sector health care

Name	Med Express Urgent Care
Address	8702 University Blvd
City	Coraopolis
State	PA
ZIP Code	15108
Website	http://www.medexpress.com/local-centers/pennsylvania/greater-pittsburgh-area/moon-township.aspx

3. Drag the pop-up away from your map.

4. Point to each of the buttons on the lower right of the pop-up window to read what they do, and click the Zoom To This Feature button several times. The map centers and zooms in on the

Coraopolis MedExpress Clinic. If you zoom in close enough, the buffers and population density displays turn off, the streets display turns on, and the MedExpress clinic is labeled. If you zoom in even farther, the streets are labeled. These feature classes and labels have visibility ranges for which they are visible.

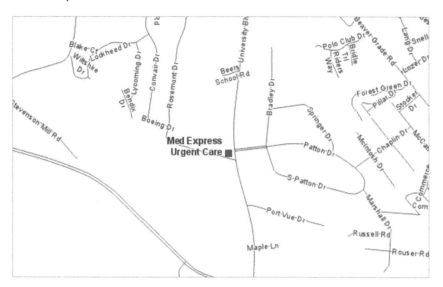

5. **Close the pop-up, and zoom to full extent.**

Zoom in

1. **Position the pointer over the point where the three rivers join in the middle of the map, and use the wheel button to zoom in on the map, stopping several times before zooming in more.** You can also press the plus key (+) to zoom in and minus key (–) to zoom out.

2. **Scroll up and down in the Contents pane.** Feature classes not drawing at this scale have gray check marks, whereas feature classes that display have black check marks. You'll learn how to set visibility thresholds in chapter 3 for controlling the scale at which feature classes display.

3. **Click and pan the map to a new location.** Your pointer is automatically in panning mode with the Explore button selected in the Navigate group on the Map tab of the ribbon. You can also use the arrow keys to move about (pan) the map. If you have a touch screen, tap and slide the map to pan.

4. **On the Map tab, in the Navigate group, click the Previous Extent button** ← **a couple of times.** Clicking this button moves you back through the sequence of zooming steps you have taken. There's also a Next Extent button → for the other direction.

5. **Zoom to full extent.**

Zoom in to a raster feature class

All but one of the map layers in the Health Care Clinics map, Poverty Index, have vector data, made up of points (for example, MedExpress Clinics), lines (streets), or polygons (closed areas such as the census tracts that display population density). GIS does not store images of vector feature classes but instead draws them on the fly from stored points and drawing instructions, including how the map designer wants them symbolized (such as the brown square symbols for MedExpress clinics). Rasters, however, are stored using image data formats (for example, JPEG and TIFF) and rendered pixel by pixel as stored. However, as a map designer, you can change the colors of certain kinds of rasters, such as the Poverty Index that you are about to use.

1. **Turn on and off feature classes so that the following feature classes are on: Tract Centroids (center points), Pittsburgh, Allegheny County, Streets, and Poverty Index.**

    Hint: Press Ctrl and click a check box for a feature class to turn all feature classes on or off. Then adjust for the desired map.

    The raster feature class, which you will create in chapter 10, is rectangular, as are all raster feature classes. In this case, the raster boundaries were defined by the Track Centroid points that are farthest to the north, east, south, and west.

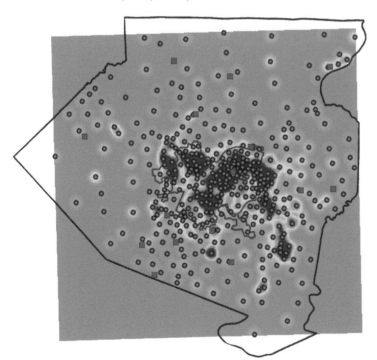

2. **Click to clear Tract Centroids. Zoom in to the center of the map until you can see the pixels of Poverty Index.**

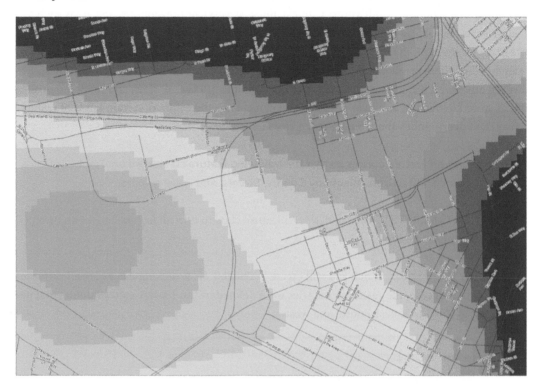

3. **Zoom to full extent, and turn on all feature classes except Tract Centroids, Municipalities, and Poverty Index.**

Use bookmarks

Spatial bookmarks allow you to zoom to preset map views.

1. **On the Map tab in the Navigate group, click the Bookmarks button** 📖. Three bookmarks are available for the open map: Allegheny County, Poverty Areas, and Pittsburgh East End. Also available are three bookmarks for a 3D map that you'll use in tutorial 1-4.

2. **Click the Poverty Areas bookmark.** The map zooms to that area. Your map may have different feature classes displayed or not displayed, depending on the size of your computer's screen and map window, because the thresholds that switch feature class displays on and off depend on the ratio of feature sizes on the screen to actual feature sizes on the ground. You will learn more about map scale in chapter 3.

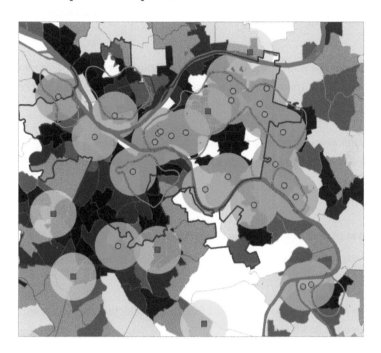

3. **In the lower-right corner of the graphic, zoom and pan in to the poverty risk area until streets appear.** An alternative is to press the Shift key, and drag a rectangle around the area desired for viewing.

4. **On the Map tab, in the Navigate group, click Bookmarks > New Bookmark.**

5. **In the Create Bookmark window, type** McKeesport Poverty Area **for Name, and click OK.**

6. **Click the Allegheny County bookmark, and then try out your new bookmark.**

7. **Click Bookmarks > Manage Bookmarks.**

8. **In the Bookmarks pane, alphabetize the Health Care Clinics bookmarks by dragging them in order.**

9. **Close the Bookmarks pane.**

10. **Zoom to full extent.**

Search for a feature

Next, you will use the ArcGIS Pro query builder for Structured Query Language (SQL) queries. SQL is the standard language for querying tabular data. In this quick preview, you'll search for locations on the basis of their attribute data values. Chapter 4 reviews SQL search criteria in more detail.

1. In the Contents pane, clear the Population Density check box, and select the Municipalities check box.

2. **Right-click Municipalities to open the menu, and click Attribute Table.** Every vector feature class has an attribute table, and each feature (point, line, or polygon) of a feature class has a record or row of data.

3. **On the Map tab, in the Selection group, click the Select By Attributes button** . The Geoprocessing pane appears. Municipalities is already selected as the input feature class.

4. **Click the New Expression button.**

5. **Select Name, is equal to, and McKees Rocks from the drop-down lists.**

6. **Click OK.** The result is that the McKees Rocks record and feature are selected. You will complete the next two steps to see the record and feature.

7. **At the bottom of the Municipalities table, click the Show Selected Records button** ▥.

8. **In the Contents pane, right-click Municipalities, point to Selection, and click Zoom To Selection.**

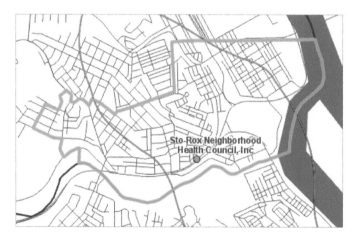

9. **At the top of the table, click the Clear button** ▤.

10. **Clear the Municipalities check box, and check the Population Density check box.**

YOUR TURN

Search for an FQHC, the Braddock Health Center, and zoom to the health center. When you finish, close the Geoprocessing window and any open tables, clear selections, and zoom to full extent. Save your project.

Tutorial 1-3: Work with attribute data

Attributes play a major role in GIS. Besides providing data needed to solve a problem or investigate spatial patterns, attributes allow you to search for useful information and mapped features, as seen in the previous tutorial. Attributes also enable sophisticated symbolization and labeling, as you'll see later in this chapter.

Open the Tutorial 1-3 project

1. **Open and save Tutorial1-3.aprx as** Tutorial1-3YourName.aprx.

2. **Zoom to full extent.**

Open and sort attribute tables

You'll start with a closer look at feature attribute tables.

1. **In the Contents pane, right-click MedExpress Clinics, and click Attribute Table.** The table shows 15 MedExpress clinics with name and address data available, along with latitude and longitude coordinates, hyperlink URLs, and other information.

OBJECTID	Shape	Name	Address	City	State	ZIP Code	Latitude	Longitude	Website
1	Point	Med Express	4655 William Flynn H...	Allison Park	PA	15101	40.591454	-79.948107	http://www.medexpress.com/local...
2	Point	Med Express	3024 Washington Pike	Bridgeville	PA	15017	40.353313	-80.115109	http://www.medexpress.com/abo...
3	Point	Med Express	2644 Mosside Blvd #...	Monroeville	PA	15146	40.43207	-79.751309	http://www.medexpress.com/local...
4	Point	Med Express	1535 Washington Rd	Pittsburgh	PA	15228	40.357001	-80.050376	<Null>
5	Point	Med Express Urgent...	8702 University Blvd	Coraopolis	PA	15108	40.506133	-80.223608	http://www.medexpress.com/local...
6	Point	Med Express Urgent...	2600 Old Washingto...	Pittsburgh	PA	15241	40.310486	-80.090711	http://www.medexpress.com/local...

2. **If necessary, adjust the column widths in the table so that you can read the full cell contents by positioning the pointer between column names on the top row until the pointer becomes a two-headed arrow, and click and adjust by moving left or right.** You can also double-click when you see the two-headed arrow to automatically resize the column widths.

3. **In the table, drag the Website column after ZIP Code.**

4. **Right-click the City column heading, and click Sort Ascending.** Now, records are sorted by city name. If you scroll down, you can see that eight of the MedExpress clinics are in Pittsburgh.

5. **Right-click City, and click Custom Sort.** With this option, you can sort more than one column. Next, you'll sort by city and an address within a city.

6. **In the Custom Sort window, for Field, select City and Address.**

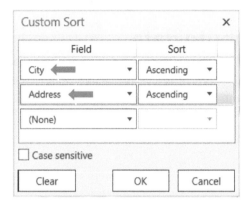

7. **Click OK, and scroll down to the Pittsburgh records.** If there were many records, you could now easily find a Pittsburgh MedExpress by address because street numbers are in order, sorted as text. Note that numbers stored and sorted as text are ordered by individual digits instead of numeric value. For example, 11 appears before 2 because the first 1 of 11 is smaller than the 2.

4	Point	Med Express	1535 Washington Rd	Pittsburgh
13	Point	Med Express Urgent C...	1984 Greentree Rd	Pittsburgh
6	Point	Med Express Urgent C...	2600 Old Washington...	Pittsburgh
16	Point	Med Express Urgent C...	3433 William Penn Hwy	Pittsburgh
18	Point	Med Express Urgent C...	3516 Saw Mill Run Blvd	Pittsburgh
23	Point	Med Express Urgent C...	5201 Baum Blvd	Pittsburgh
9	Point	Med Express Urgent C...	695 Clairton Blvd	Pittsburgh
26	Point	Med Express Urgent C...	7219 Mcknight Rd	Pittsburgh

8. **Close the MedExpress Clinics table, and save the project.**

YOUR TURN

Open the attribute table for Population Density. GeoID is a geocode (unique identifier or primary key) assigned to census tracts by the US Census Bureau. Attributes of interest are Pop (2010 population), Area (square miles), and PopDensity, which is Pop/Area (persons per square mile). Using sorting, find the tract with the highest population density, 29,835 persons per square mile. Select that record (click the gray square on its left), right-click and select Zoom to, and find it on the map (cyan area in the middle of the map). To get a better look at the selected census tract, zoom out, to the extent that Population Density's display stays on. Click the Clear button at the top middle of the table. Close the table when finished, and zoom to full extent.

Work with the field view of an attribute table

You can change the order of attributes (columns) in a table, change the names and displayed names (aliases) of attributes, see the data type of attributes, and make only certain attributes visible to the user—all using the Fields view of a table.

1. **In the Contents pane, turn on Tract Centroids.**

2. **Open this feature class's attribute table.** Four attributes of this feature class are indicators of poverty: FemHseHld is the number of female-headed households with children, Unemp is the number of unemployed persons age 16 and older who are in the workforce, PopNHighSc is the population age 24 or older that did not attain a high school education, and PopPov is the number of persons below the income poverty level.

3. **In the upper-right corner of the table, click the Options button ☰, and click Fields View.** A new tab opens with the fields of Tract Centroids and their properties.

☑ Visible	◼ Read Only	Field Name	Alias	Data Type	☑ Allow NULL	☐ Highlight	Number Format	Domain	Default	Length
☑	☑	OBJECTID	OBJECTID	Object ID	☐	☐	Numeric			
☑	☐	Shape	Shape	Geometry	☑	☐				
☑	☐	GEOID	GEOID	Text	☑	☐				11
☑	☐	FemHseHld	FemHseHld	Long	☑	☐	Numeric			
☑	☐	Unemp	Unemp	Double	☑	☐	Numeric			

4. **Type the following aliases:**
   - **For FemHseHld, add the alias** Female Headed Households.
   - **For Unemp, add the alias** Unemployed Population.
   - **For PopNHighSc, add the alias** High School Non-Graduates.
   - **For PopPov, add the alias** Poverty Population.

5. **Drag the PopPov row to just below GeoID.**

6. On the Fields tab, in the Changes group, click the Save button, and close the Fields view. The Tract Centroids table reflects the changes you just made.

OBJECTID	Shape	GEOID	Poverty Population	Female Headed Households	Unemployed Population
1	Point	42003402000	529	297	80
2	Point	42003423000	233	113	67
3	Point	42003270400	365	162	15
4	Point	42003270800	323	162	143
5	Point	42003426700	180	75	94

7. Close the Tract Centroids table, and turn off its feature class.

YOUR TURN

In the Municipalities table, turn off visibility for all attributes except Name. Save your changes, and close the Fields view and the table. Note that to publish this map to ArcGIS Online, you must select the visibility check box for Shape.

Select records and features of a map feature class

1. Open the attribute table of FQHC Clinics.

2. In the table on the left of row 1, click the square gray cell, and drag down through row 6. The result is that you have selected six of the 26 FQHC Clinics records in the table and highlighted their point symbols on the map (with the cyan selection color), demonstrating the linkage between records and features. Many GIS functions work with selected subsets of records and features.

3. At the top of the attribute table, click the Clear button.

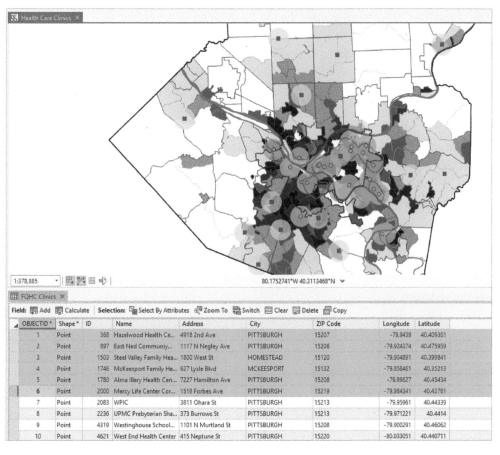

OBJECTID *	Shape *	ID	Name	Address	City	ZIP Code	Longitude	Latitude
1	Point	368	Hazelwood Health Ce...	4918 2nd Ave	PITTSBURGH	15207	-79.9439	40.409361
2	Point	897	East Ned Communiy...	1117 N Negley Ave	PITTSBURGH	15206	-79.924374	40.475959
3	Point	1503	Steel Valley Family Hea...	1800 West St	HOMESTEAD	15120	-79.904891	40.399841
4	Point	1746	McKeesport Family He...	627 Lysle Blvd	MCKEESPORT	15132	-79.858461	40.35213
5	Point	1780	Alma Illery Health Cen...	7227 Hamilton Ave	PITTSBURGH	15208	-79.89627	40.45434
6	Point	2000	Mercy Life Center Cor...	1518 Forbes Ave	PITTSBURGH	15219	-79.984341	40.43761
7	Point	2083	WPIC	3811 Ohara St	PITTSBURGH	15213	-79.95961	40.44339
8	Point	2236	UPMC Prebyterian Sha...	373 Burrows St	PITTSBURGH	15213	-79.971221	40.4414
9	Point	4319	Westinghouse School...	1101 N Murtland St	PITTSBURGH	15208	-79.900291	40.46062
10	Point	4621	West End Health Center	415 Neptune St	PITTSBURGH	15220	-80.033051	40.440711

4. **At the top of the Contents pane, click the List By Selection button** .

5. **Clear the check boxes as needed so that only the MedExpress and FQHC Clinics are turned on.**
Now when you click the map, you can select only clinics and not mistakenly select other features.

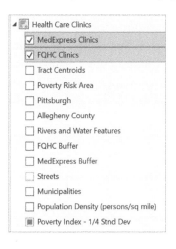

6. **At the top of the Contents pane, click the List By Drawing Order button** ⬚.

7. **On the Map tab, in the Selection group, click the Select button** ⬚.

8. **Press the Shift key, and on the map, individually select any five FQHCs.** You will see five corresponding records selected in the table.

9. **At the bottom left of the table, click the Show Selected Records button.** Now you will see only the selected records.

10. **Click the Show All Records button** ⬚.

11. **At the top of the table, click the Switch Selection button** ⬚ Switch. Now all the records except for the five FQHCs that were originally selected are now selected.

12. **Clear the selection.**

YOUR TURN

Using the Selection tool, drag a rectangle around some FQHCs on the map. All FQHCs within the rectangle are selected. Press and hold the Shift key, and drag a different rectangle. More FQHCs are added. Press and hold the Ctrl key, and click an already selected FQHC. That FQHC gets deselected while all other selected FQHCs remain selected. Press and hold the Shift key, and reselect the FQHC you just deselected. That FQHC gets added back to the selection. Clear the selections, and close the table. This assignment showed that you can select any subset of features.

Get summary statistics using a tool

ArcGIS has hundreds of tools, each with inputs and algorithms that transform the inputs into outputs. You can search for a tool, fill out its form to specify inputs, set parameters that control algorithm behavior, and name and specify where to store outputs. The Summary Statistics tool computes common statistics (for example, minimum, maximum, mean, and standard deviation) and writes results to a new table. Obtaining and studying summary statistics for attributes of interest are among the first steps of any analysis.

1. **On the Analysis tab, click the Tools button** ⬚.

2. **In the Geoprocessing pane, click the Toolboxes link.**

3. **Expand Analysis Tools > Statistics.**

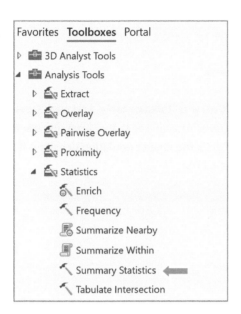

4.   **Click the Summary Statistics tool.**

5.   **In the Summary Statistics tool dialog box, choose Population Density (persons/sq mile) and set the Statistics Type drop-down menu options to Minimum, Maximum, Mean, and Standard deviation to get the statistics for PopDensity of the Population Density feature class. Type** AllCoTractsStatistics **for Output Table.** ArcGIS Pro automatically chooses your project's default file geodatabase, Chapter 1.gdb, for the output table, AllCoTractsStatistics. Make sure that the Case field is clear. If it's not clear, point to the left of the field and click the delete button.

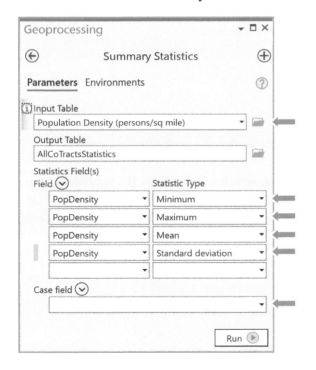

6.   **Click the Run button.**

7.   **When the run completes, close the Geoprocessing pane.**

8.   **In the Contents pane, scroll down, right-click the AllCoTractsStatistics table, and click Open.** You will find the results in the table as shown in the figure. You can see that at least one tract has no population, the maximum population density is 29,835, the mean population density is 4,628 persons per square mile, and the standard deviation is 4,110.

OBJECTID	FREQUENCY	MIN_PopDensity	MAX_PopDensity	MEAN_PopDensity	STD_PopDensity
1	402	0	29835.128906	4628.170052	4110.425256

9.   **Close the table, and save your project.**

Tutorial 1-4: Symbolize maps

This tutorial introduces you to symbolizing maps. You'll change the point symbol of feature classes, including type of symbol, color, and size. You'll label features with their name, choose a font and size, and place a halo around labels to improve readability. You'll add map feature classes to the map from your hard drive, symbolize them, and remove them from your map. Finally, you'll add aerial imagery to your map from ArcGIS Online as a map service. Chapter 2 goes into depth on symbolization.

Open the Tutorial 1-4 project

1.   **Open and save Tutorial1-4.aprx as** Tutorial1-4YourName.aprx.

2.   **Use the Allegheny County bookmark.**

Symbolize feature classes

Now examine how ArcGIS Pro allows you to choose symbols for vector feature classes.

1.   **In the Contents pane, right-click FQHC Clinics, and click Symbology.** The Symbology pane appears. Here you see that symbology is Single Symbol, the current symbol is a green circle with a dark-gray boundary, and you see fields for typing a new label and description for the Contents pane and layout legend if desired.

2.   **In the Symbology pane, click the current symbol (the green circle).** A gallery of symbols opens.

3.   **Click Circle 4.** The FQHC symbols on the map immediately change to Circle 4.

4.   **At the top of the Symbology pane, click Properties.**

5. In the Symbology pane under Appearance, change the color to Leaf Green (seventh column, fifth row), change size to 8 pt, and click Apply.

YOUR TURN

Symbolize Poverty Risk Area with a 1.5 pt line and Poinsettia red color (second column, fourth row). Close the Symbology pane when finished.

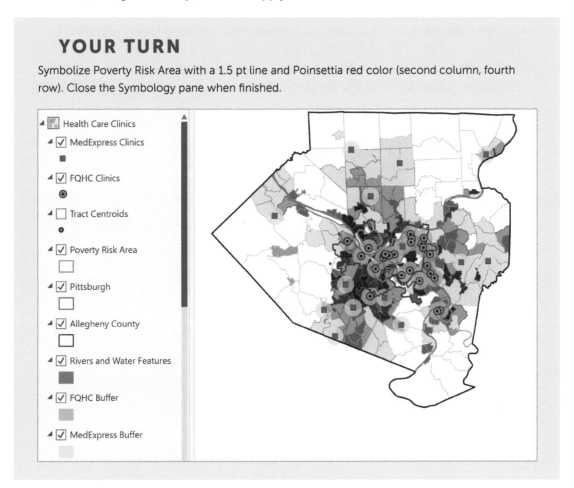

Label a feature class

Next, you will label a feature class using one of its attributes.

1. In the Contents pane, turn off Population Density, and turn on Municipalities.

2. Click Municipalities. Because you selected this feature class, its Contextual tab, Feature layer, appears on the ribbon to give you access to functionality for modifying or enhancing the feature class. If you don't see the Feature layer tab, widen your ArcGIS Pro window.

3. Under Feature layer, click Labeling.

4. Click the Label button ✎ to turn labeling on for Municipalities. Municipalities are automatically labeled with their Name attribute. Next, you'll make the labels less prominent.

5. **In the Text Symbol group, change the font to size** 7 **and select a dark-gray color. If you don't see** Text Symbol, make the ArcGIS Pro window wider.

6. **In the Text Symbol Group, in the lower-right corner, click the Dialog Launcher button** 🖼️.

7. **In the Label Class pane, click Symbol, scroll down if necessary, click Halo. For Halo Symbol, click the white square for White Fill. For Outline Color, choose No Color, and for Halo Size, type** 0.75 **pt.**

8. **Click Apply, and close the Label Class pane.** In effect, the halo erases nearby features so that the full label is easy to read.

9. **Zoom in to the western part of the county that's centered on the river there.**

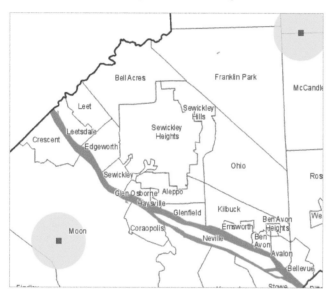

10. **Turn Municipalities off, turn Population Density on, and zoom to full extent.**

Add and remove feature classes

1. **On the Map tab, in the Layer group, click the Add Data button** ➕. Note that under Project in the left panel of the Add Data window is Folders, which provides direct links to certain folders.

2. **Click Folders to see connections to the Chapter1 folder and the Tutorials (Chapter1\Tutorials) folders.** Some chapters have a Data (Chapter1\Data) folder, from which you will need to add data to your map. In that case, use the corresponding Chapterx (where x is 1, 2, and so on) folder connection. Most times, however, you can use the Tutorials connection, which gives you direct access to the default file geodatabase—in this case, Chapter1.gdb. Note that if your project does

not have a useful connection, you can always use the Computer portion of the left panel of the Add Data pane and browse in the C or other letter drive to the location of the needed data.

3. **Double-click the Tutorials connection, double-click Chapter1.gdb, and double-click Parks.** Doing that adds Parks to the Contents pane of your map and displays the layer with an arbitrary color fill.

4. **In the Contents pane, click the Parks symbol (rectangle under Parks) to open the Symbology pane.**

5. **In the Symbology pane, under Gallery, scroll down, and click the Park symbol (light-green with no boundary).**

6. **Close the Symbology pane.**

7. **Turn off the buffer and Population Density feature classes to get a better look at Parks.**

8. **Use the Pittsburgh East End bookmark, and zoom in more if streets do not appear.** Most often you can symbolize physical features, such as rivers and parks, with colors that you'd expect for them—for example, blue for water and green for features such as wooded areas or parks.

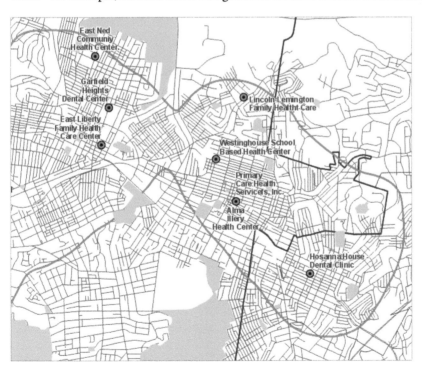

9. **In the Contents pane, right-click Parks, and click Remove.**

10. **Zoom to full extent, and save your project.**

Arrange 2D and 3D maps side by side

The Tutorial 1-4 project file has a 3D version of Population Density, with population density extruded vertically by census tract that you'll use next. The variation in population density in an urban area is so wide that it's difficult to appreciate its variation with color coding alone in a choropleth map. In 3D, the differences are impressive. In this exercise, you'll view the 2D and 3D maps of Population Density side by side and will be able to appreciate the added information of the 3D map over the 2D map.

1. **Turn on only the Allegheny County and Population Density layers.**

2. **In the Catalog pane, expand Maps, double-click Health Care Clinics_3D to open the 3D map, and close the Catalog pane.**

3. **Right-click the Health Care Clinics_3D tab, and click New Vertical Tab Group.** Both the 2D and 3D maps appear in separate, same-size windows.

4. **On the View tab, in the Link group, click the Link Views button. Click View, click Link Views, and confirm that Center and Scale is highlighted.** Now when the 2D and 3D maps are displayed side by side, you can pan, zoom, or tilt one map, and the other map will follow in the same way. If your two maps are not yet synchronized, resize and move one of the maps.

Navigate 2D and 3D maps side by side

1. **Place the pointer at the center of the 3D map, hold the wheel button down, and move the pointer straight up a bit to tilt the map.** Now you can see the tremendous differences in population density using the 3D map. You can also press V + arrow keys to tilt and rotate the 3D map. Furthermore, using the View Help button (question mark at the upper right of the ArcGIS Pro window), search for "keyboard shortcuts" and view the help document, "Keyboard shortcuts for navigation—ArcGIS Pro."

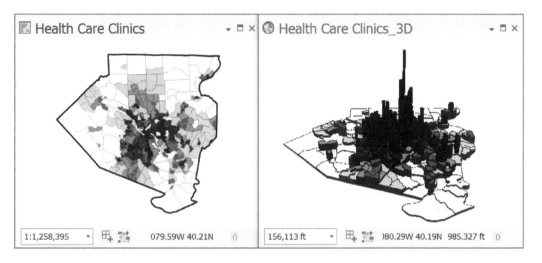

2. **Zoom and pan with either map.** Both maps move together.

3. **Position the pointer on the bottom of the 3D map, hold down the wheel button, and move your mouse to the right to rotate the 3D map (or use V + arrow keys).**

4. **Keep experimenting, panning and rotating the map to any viewing.** The figure shows the 3D map viewed from the north.

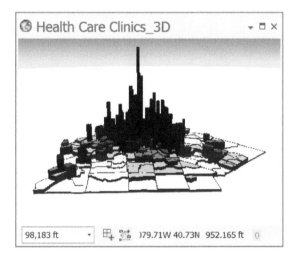

5. **Press the N key on your keyboard to reorient the map with north up.**

6. **Save your project.**

Tutorial 1-5: Publishing maps to ArcGIS Online

Generally, you'll use ArcGIS Pro as your map authoring package. With feature classes prepared and added, symbolization completed, and so on, you can share your interactive map. You can click a button and publish your map to ArcGIS Online. Once your map is online, you can modify its configuration as well as use many map navigation and query tools. You can keep your map private or share it with a group or publicly for anyone to view.

Share your map online

Feature classes for publication benefit from having a certain coordinate system (projection)—WGS84 Web Mercator Auxiliary Sphere—which is the one used in ArcGIS Online. The maps for sharing in this chapter all have the Web Mercator coordinates. Using this projection guarantees that feature classes will work online. Chapter 5 reviews map coordinates and projections.

1. **Open Tutorial1-5**. The map has only essential feature classes for sharing. The original Streets feature class stored on your computer is not included because it has a large file size. Instead of uploading and adding that feature class to your Esri account, street features are now viewable via the Streets basemap provided by an Esri web server. When you zoom in, more detailed streets appear.

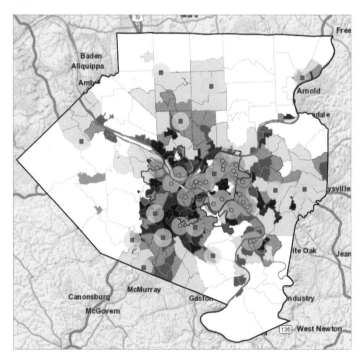

2. **Right-click MedExpress Clinics, and click Properties.**

3. In the Layer Properties: MedExpress Clinics window, click Source. In the table under Data Source, you will see the Database row. The source is a feature class, MedExpress, stored in Chapter1.gdb, a file geodatabase on your computer. You'll upload MedExpress and other feature classes to your account in ArcGIS Online.

4. **Scroll down and expand Spatial Reference to see that the feature class's projection is WGS84 Web Mercator Auxiliary Sphere.**

5. **Click Cancel.**

6. **Using steps 2 through 4, look at the source for World Street Map.** An Esri website provides the ArcGIS map service. The map service projection is Web Mercator.

7. **On the Share tab, in the Share As group, click the Web Map button** 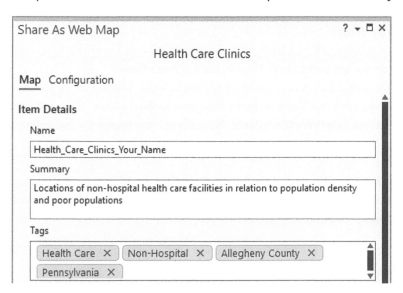.

8. **In the Share As Web Map pane, type the underscore characters and replace Your_Name with your name or student ID, as shown in the figure: Health_Care_Clinics_Your_Name.** This form is already partially filled in, but in the future with new maps, you'll have to fill in the form yourself. Note that if you are in a class sharing an organizational account, all web maps published to ArcGIS Online must have unique names. So if another person in your class has the same name, add your middle initial or middle name so that your item name is unique.

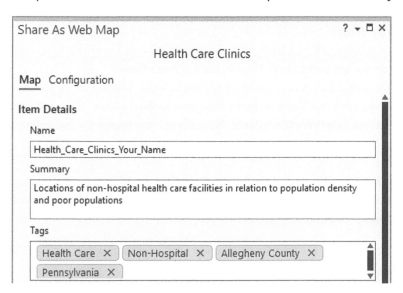

9. **In the Share As Web Map pane, click Analyze.**

10. **Click Share, and after the web map is published, close the Share As Web Map pane.** Because you are signed in to your ArcGIS account, ArcGIS Pro knows where to publish your map and web layers—namely, to the Content folder of your ArcGIS Online account. Your map may take a few minutes to publish. When ArcGIS Pro finishes, you will get a message that you successfully shared your web map.

11. **On the Project tab, save your project and click Exit.**

Open your map in ArcGIS Online

1. **In your browser, search for** ArcGIS Online.

2. **Sign in with your ArcGIS organizational account, click Content, and click My Content.** You'll see your published web map, Health_Care_Clinics, with your name added at the end.

☐ 🗺️ Health_Care_Clinics_Your_Name	Web Map
☐ 🗺️ Health_Care_Clinics_Your_Name_WFL1	Feature Layer (hosted)
☐ 📄 Health_Care_Clinics_Your_Name_WFL1	Service Definition

Share and open your map

1. **Click the check boxes on the left of the three Health_Care_Clinics items.**

2. **At the top of the list of files, click the Share button > Everyone (public) > Save.** Notice that the Shared column of the list now has Everyone entered. Now anyone can search for and use your map.

3. **Click the More Options button at the end of the Health Care Clinics web map, and click Open in Map Viewer.** Your interactive map, Health_Care_Clinics, opens in your browser. Each published feature class is now called a feature layer, or simply a layer.

Review functionality available for feature layers

1. **Click on the MedExpress Clinics layer.** The menu for this layer appears on the right of the map (if the menu does not appear, click the expand button << in the lower right of the Map Viewer window). Options include Properties, Styles, Filter, Configure Pop-Ups, and so on. Properties is selected and open in the pane on the left of the layer menu.

2. Click the Styles button. Doing this opens a pane that provides you with options for symbology. For now, you will not make any changes to this layer's symbology.

3. Click Done to collapse the menu.

Rename a layer

1. In the Layers pane, click the three dots at the end of MedExpress Clinics, and click Rename.

2. Rename the feature class MedExpress Urgent Care, and click OK.

Disable unnecessary pop-ups

When you click a location on the map, pop-ups appear with data about all the features at that location. Pop-ups are useful for some layers, but not others. By default, all layers have pop-ups enabled, so next you will disable pop-ups for layers generally not of interest in regard to their data.

1. In the Layers pane, click Poverty Risk Area.

2. Click the expand button at the lower right of the map viewer to expand the menu for Poverty Risk Area.

3. Click Configure pop-ups in this layer's menu, and in the Pop-ups pane, click the currently active Enable pop-ups button to disable pop-ups.

4. Likewise, disable pop-ups for all remaining layers except MedExpress Urgent Care and FQHC Clinics.

Configure pop-ups

1. In the Layers pane, click MedExpress Urgent Care.

2. In the Configure Pop-up pane, click Fields, and click Select Fields.

3. In the Select Fields pane, add check marks as needed for all rows except ObjectID, and click Done.

4. In the Fields list, hover over the six dots in the Name attribute, and click to drag it to the top of the list.

5. Move Website to the bottom of the list.

6. **In the layer menu on the right, click Configure Fields.**

7. **Select Latitude, change the Significant digits to** 6 **Decimal places, and click Done. Do the same for Longitude.** Note that the Save button in the left panel has a blue dot, meaning that there are unsaved changes.

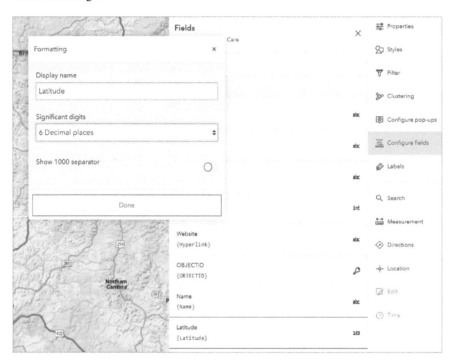

8. **In the left panel, click the Save button, and select Save.**

Use a pop-up

1. **Click the westernmost MedExpress Urgent Care facility to see its pop-up.**

2. **In the pop-up, next to Website, click View.** The MedExpress website opens.

3. **When you finish exploring the website, close the MedExpress web page and the pop-up.**

> ## YOUR TURN
>
> Configure pop-ups for the FQHC Clinics feature class using the previous directions. Select all fields except OBJECTID and ID. This layer does not have a Website field, so skip the step referencing that field. Try out a pop-up for an FQHC in the center of the map. Save your map when finished.

Change style for a point layer

1. In the Layers pane, click MedExpress Urgent Care, and in its menu, click Styles.

2. Under Try A Drawing Style, click Style Options.

3. In the Style Options pane, click the current symbol.

4. Click the square symbol, change the size to 6, and for colors, select orange (second from the left in the top row).

5. In the Symbol pane, scroll down if necessary, and at the bottom, click Outline.

6. Click Stroke, and make the width 1.5.

7. Click Done, and in the Styles menu, click Done.

Change style for a polygon layer

In this exercise, you will change the outline color of the Poverty Risk Area polygon layer from red to a darker red. You will use the red, blue, green color scheme for specifying a custom color. Color values of this scheme range from 0 to 255. The current red color of the Poverty Risk Area polygon outlines is (255, 0, 0), meaning full red and no blue or green. You'll change the value to (210, 0, 0), making the red darker.

1. In the Layers pane, click Poverty Risk Area, and in its menu, click Styles.

2. For Try A Drawing Style, click Style Options.

3. Click the current Symbol style (shown as red outline polygon) > Outline > Custom color. Note that as you set a color, the polygon in Symbol style changes to the set color, so you can see if you like it.

4. Set the Red value to 210 (you can type in the value box), click Done, and click Done again. Now you can see your darker red color for Poverty Risk Area outlines on the map.

Label a feature layer

1. Using the Zoom In button on your map repeatedly (or you can use your mouse wheel button), zoom in to the center of the map so that the FQHC Buffer, MedExpress Buffer, and Population Density layers turn off in the map display. This action allows you to see the streets basemap

within Pittsburgh's boundaries. You can drag the map as needed by clicking, holding, and dragging.

2. **In the Layer pane, select FQHC Clinics, and click Labels.**

3. **In the Label Features pane, confirm that Enable Labels is on.**

4. **Click the More Options button (three dots) in the Default label class, and click Delete.**

5. **Click Add Label Class, and click the Edit button next to Edit Label Style.**

6. **Change the font to Arial Bold Italic, size to 12, color to light green (fifth from the left in top row), halo color to white, and halo size to 2, and close the Label style pane.**

7. **In the Label features pane, slide the left side of the Visible range to the current map scale indicated by the solid black triangle.** As a result, the MedExpress Urgent Care labels appear only at the current scale or zoomed in further.

8. **Use the Zoom Out button on your map to zoom out once.** Your labels turn off.

9. **Click the Default Map View button (above the Zoom In and Zoom Out buttons).**

YOUR TURN

Add a label of your own design for MedExpress Urgent Care.

Show an attribute table

1. **In the Contents pane, click Population Density and click Show Table.**

2. **In the table, click the three dots after the PopDensity field and click Sort Descending.**

Use your map

1. **In the left panel, click the Legend button** 📇. Notice how the legend item display changes as you zoom in and out of the map. Only the feature layers turned on at any given map scale are included in the legend.

2. **Click the Bookmarks button, and use the McKeesport Poverty Area bookmark.**

3. Zoom further in to the McKeesport Family Health Center FQHC along Lysle Boulevard.

4. Click the Basemap button , and select Imagery.

5. Change the basemap back to Streets, and zoom to the default extent.

6. Save your map, and close your browser.

Tutorial 1-6: Use ArcGIS Explorer on your tablet or smartphone

Suppose that you have prepared the map with FQHCs for a nonprofit organization that promotes inexpensive health care for disadvantaged populations in Allegheny County and that now you must provide the map to staff members of all FQHCs. How would you get the map to them in interactive form? If you have an email distribution list for FQHC staff members, you could send a link via email to use or install the free ArcGIS Explorer app as a simple map viewer that you'll use in this tutorial. The app, which is available from the Apple App Store and Google Play, provides access to your ArcGIS Online maps for iOS or Android devices. The following instructions and figures reflect the installation workflow for Android devices, but iOS users will find it similar. You'll finish the tutorial by sending yourself an email with the link to your map.

Install and start using Explorer

1. Search the Apple App Store or Google Play for ArcGIS Explorer and install it on your smartphone or tablet.

2. Open Explorer, click Sign in with ArcGIS Online, and enter your ArcGIS account user name and password.

Open your map

1. Select your Health_Care_Clinics map.

2. Using usual gestures, zoom in to your map. The map has visibility ranges, so if you zoom in close enough, the Population Density eventually turns off.

    Various tools will help you change the basemap, draw, measure distances or areas, select the legend, select and clear layers to turn them on and off, use bookmarks, show your location, search, and share the map.

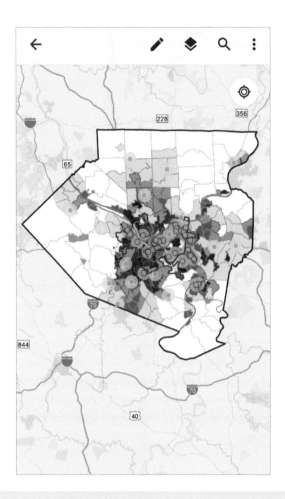

YOUR TURN

Try out all the buttons just described. When you get to sharing, send yourself an email with the link to the map. When you open the email, you'll have the option to open the map in Explorer, install Explorer, or open the map in ArcGIS Online. When you want to share a map, Explorer offers many options. When you finish, close Explorer.

Assignments

Now that you've successfully worked through this chapter's tutorials and Your Turn assignments, you can try out your new knowledge and skills in the assignments in ArcGIS Online, in the Learn ArcGIS organization's GIS Tutorial for ArcGIS Pro 2.8 (Esri Press) group, at go.esri.com/GISTforPro2.8Data. Two assignments and their ArcGIS projects and data for this chapter are available for download:

- **Assignment 1-1:** Analyze the change in population by county in the United States from 2000 to 2010.
- **Assignment 1-2:** Produce a neighborhood block watch crime map.

Map design

LEARNING GOALS

- Symbolize maps using qualitative attributes and labels.
- Use definition queries to create a subset of map features.
- Symbolize maps using quantitative attributes.
- Learn about 3D maps.
- Symbolize maps using graduated and proportional point symbols.
- Create normalized maps with custom scales.
- Create density maps.
- Create group layers and layer packages.

Introduction

In this chapter, you'll learn how to design and symbolize thematic maps. A thematic map strives to solve or investigate a problem, such as analyzing access to urgent health care facilities in a region, as you did in chapter 1. A thematic map consists of a subject layer or layers (the theme) placed in spatial context with other layers, such as streets and political boundaries.

Choosing map layers for a thematic map requires answering two questions:

1. What layer or layers are needed to represent the subject?
2. What spatial context layers are needed to orient the map reader to recognize locations and patterns of the subject features?

Often, the subjects of thematic maps are vector map layers (points, lines, or polygons), because such layers often have rich quantitative and qualitative attribute data that is essential for analysis. Of course, the subject can be a raster layer (in chapter 10, for example, you will create a risk-index raster map to identify poverty areas of a city, and poverty is the subject of the map). Spatial context layers can be vector, such as streets and political boundaries. These layers also can come in both raster or vector formats, including many basemap layers provided by Esri map services.

The major map design principle for thematic maps is to make the subject prominent while placing spatial context layers in the background. For example, if the subject is a map layer with points and you want to give them focus, you might give the point symbols a black boundary and a bright color. These subject features are known as "figure" and are the main composition of the map. Everything that is not figure is known as "ground." For example, if a context layer has polygons that are not the focus of the map, you might give the polygons a gray boundary and no color, thereby placing them in the background.

Symbolization is easy for vector maps because ArcGIS Pro can use attribute values to automate drawing. For example, ArcGIS Pro could draw all food pantry facilities in a city by using unique values with a square point symbol of a certain size and color. Continuing, the software can draw all soup kitchen facilities with a circle of a certain size and different color by using an attribute with type-of-facility code values (including "food pantry" and "soup kitchen").

In this chapter, you will learn to use good cartographic (symbolization) principles as you build several vector-based thematic maps.

Tutorial 2-1: Choropleth maps for qualitative attributes

Placing objects of all kinds into meaningful classes or categories is a major goal of science. Classification in tabular data is accomplished using attributes with codes that have mutually exclusive and exhaustive qualitative values. For example, a code for size could have the values "low," "medium," and "high." Any instance of the features with this code is displayed in only one of the classes (the values are mutually exclusive). Moreover, there are no more size classes (the values are exhaustive). In this tutorial, you learn how to symbolize mapped features—points, lines, and polygons—by class membership as available in code attributes.

Open the Tutorial 2-1 project

1. **Open Tutorial2-1.aprx from Chapter2\Tutorials, and save the project as** Tutorial2-1YourName.aprx. A New York City Zoning and Land Use map opens showing Neighborhoods and a light-gray raster basemap. Two other layers, ZoningLandUse and Water, are available but not visible yet. None of the vector layers are properly symbolized yet.

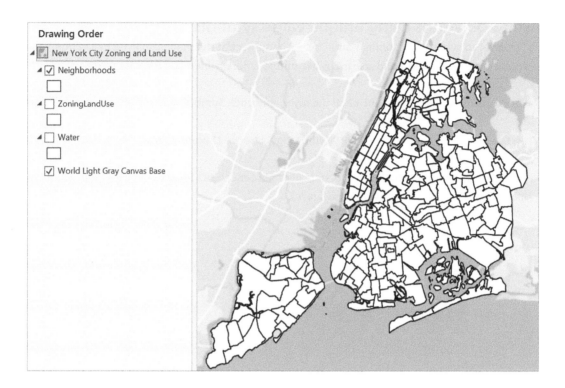

2. **Use the Lower Manhattan bookmark.** The subject of this map, zoning and land use, is best viewed and studied at approximately this zoomed scale or even closer because the geographic zones are relatively small in area. You must get close enough to distinguish them from one another.

Display polygons using a single symbol

The Neighborhoods and Water polygon layers provide spatial context. Layers should be displayed using outlines with no color fill, with water features being an exception and generally given a blue color and no outline. Context layers are easy to symbolize. You can start with Neighborhoods.

1. **In the Contents pane, under Neighborhoods, click the white color box to modify the symbol.**

2. **In the Format Polygon Symbol pane, under Properties, change Color to No Color.**

3. **Change the Outline color to Gray (60 percent), and click Apply.**

YOUR TURN

Turn on the Water layer, and symbolize the layer with a blue polygon symbol. Hint: On the Gallery tab, search for Water, and click one of the Water (area) symbols.

Display polygons using unique value symbols

The last layer to symbolize, Zoning Land Use, is the subject of the map, displayed by Unique Values on primary land-use code. Land-use maps use muted colors, which you'll create next.

1. **Turn on ZoningLandUse, right-click the layer, and click Symbology.**

2. **In the Symbology pane, for Primary symbology, choose Unique Values.**

3. **For Field 1, choose LANDUSE2.** This step adds random colors for six land uses (your colors may be different from those shown in the figure).

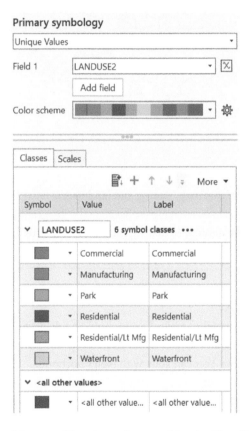

Next, you'll assign colors used by the New York City Planning Department. You'll start by changing the outlines of all polygons from black to light gray. When you view land-use polygons, their black outlines often take up too much of your map and attention and distract from the symbolized color. A gray color will soften this interference and still show boundaries.

4. **In the Symbology pane, click More > Format All Symbols.**

5. **In the Format Polygon Symbols pane, click Properties, and change Outline Color to Gray 20 percent.**

6. Click Apply, and click the back button ← to go back to the Symbology pane.

7. For Landuse2, click the symbol for Commercial, and change the color to Rose Quartz (first row, second column).

8. Click Apply, click the back button, and apply the following colors for the remaining land uses:
    *   Manufacturing: Lepidolite Lilac (first row, 11th column)
    *   Park: Apple Dust (seventh row, sixth column)
    *   Residential: Yucca Yellow (first row, fifth column)
    *   Residential/Lt Mfg.: Soapstone Dust (seventh row, third column)
    *   Waterfront: Atlantic Blue (ninth row, ninth column)

9. In the ZoningLandUse pane, click More, and clear Show All Other Values. All polygons have land-use code values, so this option is not needed. If left on, other values would be entered in the legend in the Contents pane and perhaps confuse the map reader.

10. Close the Symbology pane, and save your project. You can see all boundaries of primary land uses with their gray outlines.

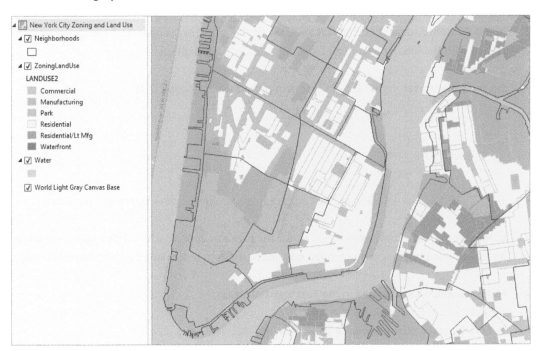

Tutorial 2-2: Labels

Labels created from attributes, such as neighborhood names, are an important part of cartography and an integral and informative component of a map. You must specify the elements of font, size, color, placement, and visibility ranges to make labels easy to read.

Open the Tutorial 2-2 project

In this exercise, you will label all three layers of the map. Each layer will have its own label properties and label placements.

1. **Open Tutorial2-2.aprx from Chapter2\Tutorials, and save the project as** Tutorial2-2YourName.aprx.

2. **Use the West Village bookmark.** To maintain visual clarity, labels for a detailed layer such as ZoningLandUse are most useful when zoomed in to the neighborhood level or similar larger scale because of the number of small polygons.

Change label properties

1. **In the Contents pane, click ZoningLandUse.** The Feature Layer contextual tab appears on the ribbon with the tabs Appearance, Labeling, and Data highlighted.

2. **On the Labeling tab, in the Label Class group, for Field, choose ZONE.**

The ZONE field has detailed zoning codes that are familiar to developers and planners.

3. **In the Layer group, click Label** ✍. Wait for the labels to appear; when they appear, the labels will be set to default values, and you must customize them to better suit the purpose of the map.

4. **Make the following changes in the Text Symbol group:**
    *   **Text Symbol Font Size: choose** 8.
    *   **Text Symbol Color: choose Gray 50 percent.**

5. **In the Visibility Range group, for Out Beyond, choose <Current>.** This step sets the visibility range for zoning labels so that they do not display when zoomed beyond the current scale. Your scale will not necessarily match the scale of the next figure, because scale varies with monitor resolution and how the application is arranged.

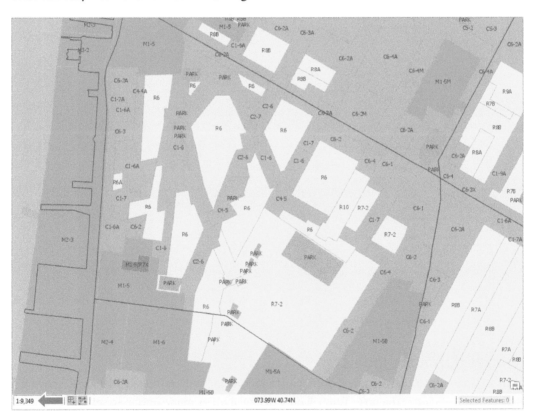

6. **Zoom in and out to see that the labels are on only when zoomed in closer than the West Village bookmark.** Note the map scale when the labels turn on or off. Depending on your screen size, the scale will vary from what is shown in the figure.

YOUR TURN

Use the Lower Manhattan bookmark. Label the Neighborhoods layer using Name, Arial font, Bold, size **7**, and a white halo. Finally, on the ribbon, on the Labeling tab in the Label Placement group, choose Land Parcel.

Label the Water layer using LANDNAME. Use the font Times New Roman, Italic, size **12**, and the color Atlantic Blue.

Set the Neighborhoods and Water labels to turn off when zoomed out beyond the Lower Manhattan bookmark. Try out the labels by zooming in and out and using bookmarks.

Remove duplicate labels

Labels for some water polygons may overlap with redundant and unnecessary labels. Removing duplicate labels will unclutter the map. You will do so using another menu option to set label properties.

1.   In the Contents pane, right-click Water > Labeling Properties.

2.   In the Label Class pane, click Position (near the top of the pane), and click the Conflict Resolution button 🖹.

3.   Expand Remove Duplicate Labels, and select Remove All.

4.   **Close the Label Class pane.** The map will now show just one label for each water feature.

5.   **Save your project.**

Tutorial 2-3: Definition queries

Often, a map layer has more features than you want to display. If so, you can use a definition query to display the desired subset of features from the larger collection, on the basis of values in the feature attribute table. For example, the point features in this tutorial start with point features for all facilities in New York City (food, health care, fire and police, schools, senior centers, and so on). You will want to display the features for food facilities only, including food pantries and soup kitchens. Defining the query allows you to select and display just these features. A definition query is different from Select By Attributes in chapter 1. The definition query is used to filter the display of a layer rather than selecting a temporary subset of features to work with, even though they both use a similar SQL interface.

Open the Tutorial 2-3 project

1.   **Open Tutorial2-3.aprx from Chapter2\Tutorials, and save the project as** Tutorial2-3YourName.aprx. An NYC Food Pantries and Soup Kitchens map opens showing Boroughs, several other spatial context layers, and the many facilities operated by the city government.

2.   **Zoom to full extent.**

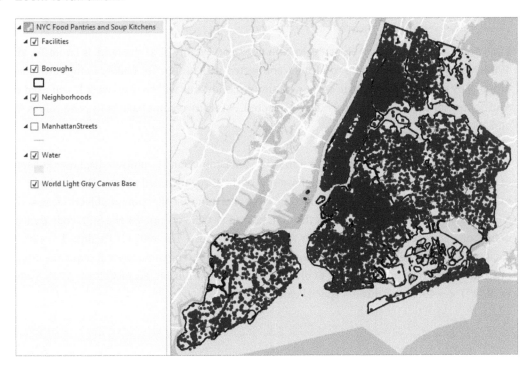

Create a definition query

In this exercise, you will create a query to display a subset of more than 20,000 government and nonprofit facilities in New York City. Facilities is a point layer of locations for services that the city provides. The map needs only three out of more than 100 classes. Classes have both a numeric code (Facility_T) and a corresponding description (Factype_1). The three facility classes needed are 4901 = Soup Kitchen, 4902 = Food Pantry, and 4903 = Joint Soup Kitchen and Food Pantry. Showing the location of these facilities could help the directors of New York City's food banks determine whether they are well located relative to poverty areas of the city.

1. In the Contents pane, right-click Facilities > Properties.

2. In the Layer Properties: Facilities window, click Definition Query > New Definition Query.

3. Select Facility_T for the Where field, Is Equal To as the logical operator, and 4901 as the value. Currently, the definition query contains a single logical condition. Only records that satisfy this condition will display.

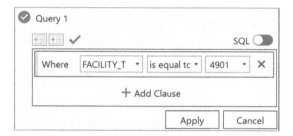

4. Click Add Clause, and select Or as the logical operator, Facility_T as the field, Is Equal To, and 4902. Now you have two single conditions connected with an "Or" to make a compound condition. Any record satisfying one of the two simple conditions will be displayed. If you used the And connector here, no records would be selected. A facility cannot have both code values 4901 and 4902.

5. Repeat step 3 with Facility_T equal to 4903. The final compound condition displays a subset of the original table of facilities, which have one of the three included values. This example extends to any definition query for selecting subsets of a finite collection of objects classified by a code such as Facility_T. If you click the SQL button SQL ⬤, you can see the SQL code that the definition query generates and is run by ArcGIS Pro: Facility_T = 4901 Or Facility_T = 4902 Or Facility_T = 4903. Because the three needed codes happen to be in numerical sequence, it's also possible to generate an alternative, equivalent compound criterion, Facility_T > = 4901 And Facility_T < = 4903, but this is a special case.

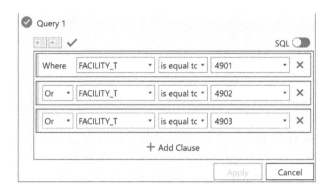

6. **Click OK.** The resulting map is a subset (631) of the original 20,000-plus facilities showing just Food Pantries, Soup Kitchens, and Joint Soup Kitchen and Food Pantries. You can verify this is true by opening the attribute table for Facilities.

Symbolize figure and ground features

The subject of the map, Food Facilities, is figure, and all other layers are ground. Figure features get accentuated with bright colors, and ground gets shades of gray.

1. **Use the Manhattan bookmark.** Someone using this map would not study the map with so many point features at full extent but would zoom in, as shown with this bookmark.

2. **In the Contents pane, click Facilities, and rename the layer** Food Facilities.

3. **In the Contents pane, right-click Food Facilities > Symbology.**

4. **In the Symbology pane, for Primary Symbology, click Unique Values.**

5. **For Field 1, choose Factype_1.** This field provides descriptions for the facility codes.

6. **Drag the Soup Kitchen value to the top, Food Pantry to the middle, and Joint Soup Kitchen and Food Pantry to the bottom.** The legend in the Contents pane will then read Soup Kitchen, Food Pantry, and Joint Soup Kitchen and Food Pantry.

    Next, you'll symbolize the three types of facilities. In this example, varying shape and color for unique point symbols is good practice. Color-blind people can use shape to identify facility classes; also facilities will remain distinguishable in black-and-white photocopies of such symbolization. The majority of people who can see color get the full effect of shape and color for seeing patterns of food facilities.

7. **Use the Symbology Gallery and Properties to change these symbols:**
    - **Soup Kitchen: Square 3, Mars Red color, size** 8 **pt.**
    - **Food Pantry: Circle 3, Cretan Blue color, size** 8 **pt.**
    - **Joint Soup Kitchen and Food Pantry: Cross 3, Solar Yellow color, size** 10 **pt.**

8. **Click More, turn off Show All Other Values, and close the Symbology pane.** Now the food facilities are sharply in figure with contrasting bright colors and shapes. For example, you can see that Manhattan has many more soup kitchens than the neighboring boroughs and that the northern part of Manhattan and the adjoining Bronx have a large cluster of food pantries.

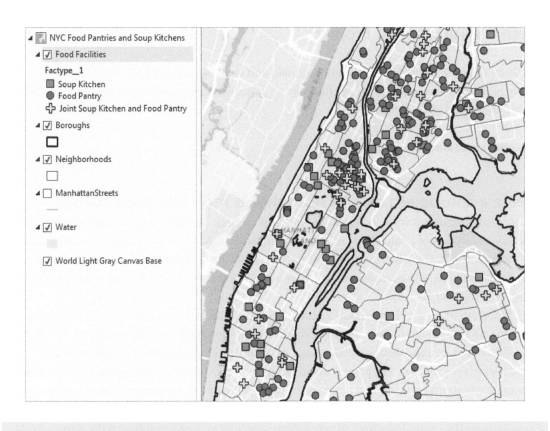

YOUR TURN

A policy decision-maker wants to know the streets where food facilities are located and doesn't want the details of a basemap. Turn off the World Light Gray basemap, and turn on Manhattan Streets. Display Manhattan Streets as a ground feature using gray (20 percent) with width **0.5** pt. Zoom to a few blocks in Manhattan and experiment with various label properties. Save your project.

Tutorial 2-4: Choropleth maps for quantitative attributes

Showing continuous variation in numerical attributes is not possible when you use the attributes to symbolize points or polygons on a map. The human eye cannot make distinctions unless there are relatively large changes in graphic elements. You must break a numeric attribute into relatively few classes (roughly three to nine), similar to how you create a bar chart for a numeric attribute. Each class has minimum and maximum attribute values. The minimum value is included in the class, but the maximum goes in the next classification to the right. To symbolize map features, you need only the set of maximum values for classes, called break points.

Open the Tutorial 2-4 project

1.   **Open Tutorial2-4.aprx from Chapter2\Tutorials, and save the project as** Tutorial2-4YourName.aprx.
     An NYC Food Stamps/SNAP Households by Neighborhood map opens showing boroughs, neigh-
     borhoods, and water features.

2.   **Zoom to full extent.**

Create a choropleth map of households receiving food stamps

A choropleth map uses color in polygons to represent numeric attribute values. Generally, increasing
color value (darkness of a color) in a color scheme represents increasing (higher) values. In this exercise, you will use US Census data aggregated to New York City neighborhoods to create choropleth
maps for households with persons over age 60 receiving food stamps/SNAP (Supplemental Nutrition
Assistance Program) benefits.

Choropleth maps use classification methods to display the data, and methods will vary depending on the data and intent of the map. The default classification method is Natural Breaks (Jenks).
This method uses an algorithm to cluster values of the numeric attribute into groups, with the
boundaries of the groups (break points) defining classes. The Natural Breaks method may be suited
for some applications in the natural sciences. However, Quantile classification is often a better
starting point, because the method is easily understood and provides information about the shape of
a distribution. The Quantile method breaks a distribution into classes—each with the same percentage of data points. For example, each quartile (quantiles with four classes) has 25 percent of the data
observations, with the middle break point being the median.

By studying quantile break points, you can determine whether a distribution is roughly uniform
(has equally spaced quantiles) or is skewed to the right (has intervals defined by break points that
become progressively larger with larger values). The former become good candidates for the Defined
Interval method (uniform distribution with easily read numbers for break points) and the latter for
the Geometric Interval method (for an increasing-width intervals distribution of break points). Many
attributes have skewed distributions.

1.   **In the Contents pane, click Neighborhoods, and rename the layer** Over age 60 receiving food
     stamps.

2.   **Open the Symbology pane for this layer, and use the following guidelines to symbolize the
     layer:**
     • **Primary symbology: Graduated Colors**
     • **Field: O60_FOOD**
     • **Method: Quantile**
     • **Classes:** 5
     • **Color scheme: Grays (5 classes)**

Primary symbology

Graduated Colors	▾

Field	O60_FOOD	▾	☒

Normalization	\<None>	▾

Method	Quantile ⟵	▾

Classes	5 ⟵	▾

Color scheme	▬▬▬▬▬	▾	⟸

Classes	Histogram	Scales

More ▾

Symbol	Upper value ▲	Label
☐ ▾	≤ 830	0 - 830
▢ ▾	≤ 1620	831 - 1620
▢ ▾	≤ 2972	1621 - 2972
▢ ▾	≤ 4831	2973 - 4831
▢ ▾	≤ 11595	4832 - 11595

3. **Click the Histogram tab.** You can see that neighborhoods with the highest quintile (quantile with five classes) tend to cluster. The interval sizes generally increase in this case, so a geometric method is an alternative to the quantile method. You will test the geometric method in Your Turn.

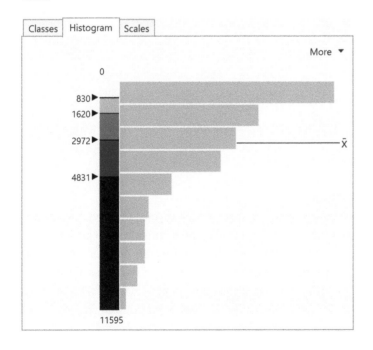

4.   The map clearly shows neighborhoods with a high number of households with persons over age 60 who receive food stamps/SNAP.

YOUR TURN

Change the symbology method for the choropleth map from Quantile to Geometric Interval. The break points become larger to provide more detail for the long tail of the distribution. The break points for the quantile method were 830, 1620, 2972, 4831, and 11595. Those for the geometric method are 881, 2201, 4181, 7148, and 11595. Reselect the Grays (5 Classes) color scheme if necessary.

Try the Defined Interval method with an interval size of **2500**. Although perhaps not the best method for this skewed data, this uniform distribution is easier to read, with multiples of 2500 at equal intervals.

Change the method back to five quantiles, close the Symbology pane, and save your project.

Extrude a 3D choropleth map

You will learn much more about 3D data and scenes in chapter 11, but you can convert a 2D choropleth map into a 3D scene to better visualize data. In particular, the map reader can get a better appreciation of extreme values relative to other values, which are not readily apparent by looking at only the color shading of choropleth maps. In 3D, features and layers are often physical features such as buildings, trees, topography, and so on, but you can display any numeric data as 3D. Here, you will learn how to convert a 2D map into a 3D scene and display neighborhood polygon features as 3D.

1. On the View tab, in the View group, click **Convert**, and click **To Local Scene**. A 3D map automatically opens with categories for 3D and 2D layers.

2. In the Contents pane, drag **Over Age 60 Receiving Food Stamps** on top of the **3D Layers** heading. Next, you will begin extruding neighborhood polygons to 3D features using the numeric attribute O60_FOOD as the extrusion height.

3. On the Feature Layer contextual tab, click the **Appearance** tab, and in the Extrusion group, click **Type**, and click **Base Height**. Clicking Base Height sets the base of the extruded polygons to zero. Next, you will set the extrusion height to the O60_FOOD attribute.

4. In the Extrusion group, for **Field**, choose **O60_FOOD**. Your features and food stamp recipient data are now displayed in 3D.

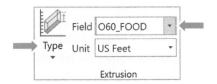

5. Scroll the wheel button on the mouse to tilt the view.

6. Zoom in to better see the extruded neighborhoods.

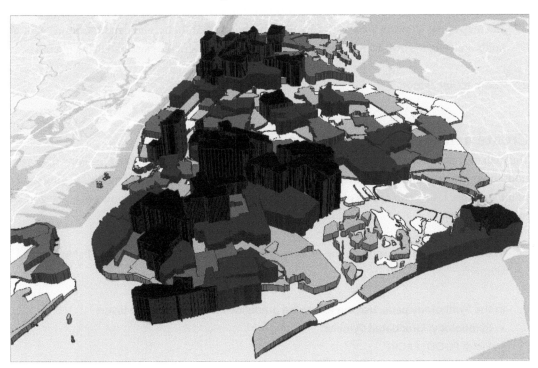

YOUR TURN

Zoom by scrolling the mouse wheel button, and pan by dragging the map. You will learn much more about 3D map navigation in chapter 11. Save your project.

Tutorial 2-5: Map using graduated and proportional point symbols

Using ArcGIS Pro, you can display polygon data using point symbols in the center (centroid) of each polygon. In the next exercise, you will create a map showing the number of food pantries and soup kitchens in New York City neighborhoods as graduated size point symbols. The larger the symbol, the more food resources in each neighborhood. You can also display data as proportional symbols that are similar to graduated symbols but represent values as unclassified symbols whose size is based on a specific value.

Open the Tutorial 2-5 project

1. Open Tutorial2-5.aprx from Chapter2\Tutorials, and save the project as Tutorial2-5YourName.aprx. The map opens showing New York City neighborhoods, boroughs, and neighborhoods with house-holds and persons over 60 receiving food stamps, which is already classified using quantiles.

2. Zoom to full extent.

Create a map of graduated size points

The map of this exercise uses two copies of the Neighborhoods polygon layer. One copy displays household data as a choropleth map. The other layer displays the number of facilities using graduated point symbols. Using two copies of the same layer is a way to show two attributes of a polygon layer in the same map.

1. In the Contents pane, rename the first Neighborhoods layer Number of food banks/soup kitchens.

2. In the Symbology pane, use the following guidelines to symbolize the layer:
   • Symbology: Graduated Symbols
   • Field: FOOD_FACIL
   • Method: Quantile
   • Classes: 5

Notice that the interval width is uniform, 2, except for the last class, so this attribute may be a good choice for the Defined Interval method (uniform distribution). An interval width of 5 is a good choice to include the maximum value of 25. Equal-width intervals are the easiest to read and are, of course, best suited for uniform distributions, but this distribution is not essential. Another possible method is defined interval, which allows you to use easily read numbers such as 1, 2, or 5 times 10 to a power (for example, 0.1, 1.0, and 10).

3. **Change Method to Defined Interval.** You may need to resize the Symbology pane to see the following options.

4. **Click the Template symbol (circle), and choose Solar Yellow as the color.**

5. **Click Apply and the back button.** Your symbology should match what is shown in the figure.

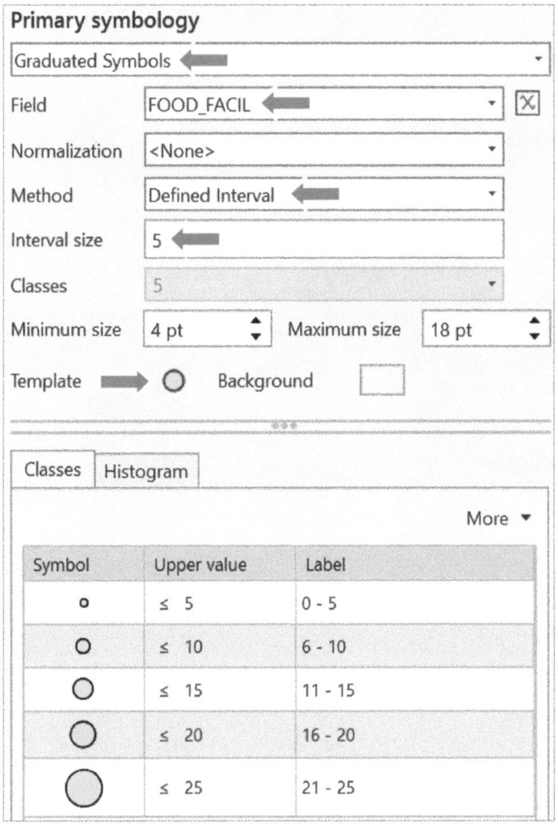

Primary symbology

Field	FOOD_FACIL
Normalization	<None>
Method	Defined Interval
Interval size	5
Classes	5
Minimum size	4 pt	Maximum size	18 pt

Template ○ Background

Classes | Histogram

More ▼

Symbol	Upper value	Label
∘	≤ 5	0 - 5
○	≤ 10	6 - 10
○	≤ 15	11 - 15
○	≤ 20	16 - 20
○	≤ 25	21 - 25

6. The finished map shows the number of food resources compared with the number of households with persons over age 60 receiving food stamps. The food pantries and soup kitchens appear concentrated in areas with poor populations, for the most part, as indicated by the map.

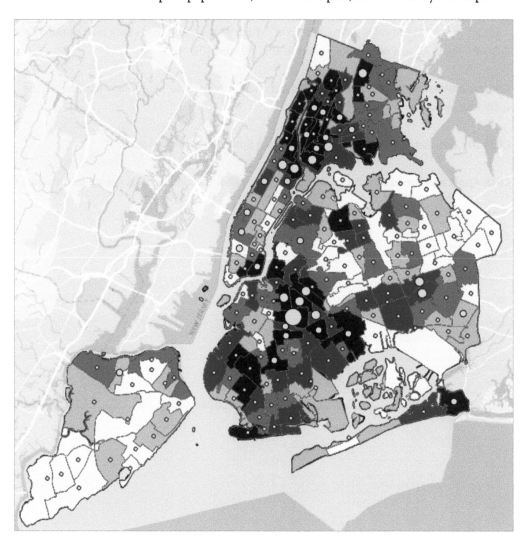

YOUR TURN

Turn on and rename the second Neighborhoods layer **Under 18 receiving food stamps**. Use Proportional Symbols, U18_FOOD as the Field, a shade of purple as the color, **2** as the Minimum size, and **20** as the Maximum size. Use the Bronx and Brooklyn bookmarks to study the relationship of food banks, soup kitchens, and persons over 60 and under 18 receiving food stamps in one map. Save your project.

Tutorial 2-6: Normalized population map with custom scales

A choropleth map showing population, such as the number of persons receiving food stamps, is useful for studying needs, such as the demand for goods and services. For example, delivery of food services for the poor requires capacities to match populations, including budgets, facilities, materials, and labor.

Choropleth maps of normalized population data have different uses than choropleth maps of populations. Dividing (normalizing) a segment of the population by the total population provides information about the makeup of areas. For example, areas with high proportions of total population receiving food stamps may be better candidates for food pantries and soup kitchens than those with low proportions, because the high-proportion areas are likely poor in many ways, including having poor geographic access to grocery stores and urgent health care.

In this tutorial, you will normalize the number of female-headed households (single mothers) with children under the age of 18 receiving food stamps by the total number of households in each neighborhood. You will find the same information for male-headed households (single fathers) with children under the age of 18 receiving food stamps and compare the two populations using a custom scale.

Open the Tutorial 2-6 project

1. **Open Tutorial2-6.aprx from Chapter2\Tutorials, and save the project as** Tutorial2-6YourName.aprx. The map opens showing New York City neighborhoods (female-and male-headed households receiving food stamps) and boroughs.

2. **Zoom to full extent.**

Create a choropleth map with normalized population and custom scale

You will create a custom classification, which often is easier to read than other classifications. The Geometric Interval method works well for representing the long tails of distributions skewed to the right, but the break points of this method do not follow a pattern that can be read easily. The custom classification of this exercise has intervals that double in width (and therefore form a geometric progression), which is read easily.

Part

1

Chapter

2

Tutorial

6

1. **Symbolize Female Headed Households Receiving Food Stamps using the following guidelines:**
    - **Symbology: Graduated Colors**
    - **Field: U18FHHFOOD**
    - **Normalization: TOT_HH**
    - **Method: Quantile**
    - **Classes:** 5
    - **Color scheme: Orange-Red (5 classes)**

    These settings show the fraction of single mothers with children under 18 receiving food stamps. Next, you will show the values as a percentage.

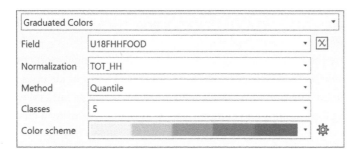

2. **Click the Symbology Options button** 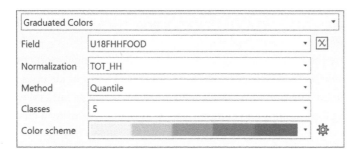**. Click Advanced Symbology Options > Format Labels.**

3. **For Category, click Percentage.**

4. **For Percentage, click Number Represents A Fraction, and for Rounding, click 0 for Decimal Places.**

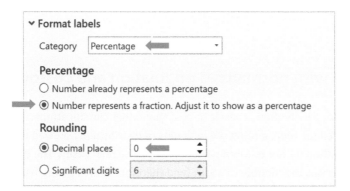

5. **Click the Primary Symbology button** **, and click the Histogram button. The** classification will now show the percentage of single mothers who receive food stamps. Expressed as percentages and rounded, the quantile break values are 1 percent, 2 percent, 6 percent, 10 percent, and 26 percent. Next, you will create custom classes using the mathematical progression 2 percent, 4 percent, 8 percent, 16 percent, and 26 percent, and the last value is the maximum.

6. **Change the method to Manual Interval.**

7. **Click the Classes tab.**

8. **In the Classes panel, click the cell for the first Upper value, type** 0.02, **and press Enter.** This makes the first class 2 percent.

9. **Continue selecting break points, and enter** 0.04, 0.08, 0.16, **and** 0.26 **(the maximum value rounded to two decimal places).**

Classes	Histogram	Scales

Symbol ▲	Upper value ▲	Label
☐ ▾	≤ 0.02	0% - 2%
☐ ▾	≤ 0.04	3% - 4%
☐ ▾	≤ 0.08	5% - 8%
☐ ▾	≤ 0.16	9% - 16%
☐ ▾	≤ 0.26	17% - 26%

10. **Click the Histogram tab.** Notice that this set of break points has wider intervals for low values than quantiles, which allows an additional value (0.16) for high values, thereby providing more information about high values.

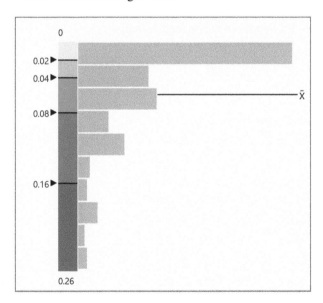

The numerical sequence of the custom break points has both increasing interval widths, as desired for the long-tailed distribution, and a recognized and easy set of values to read.

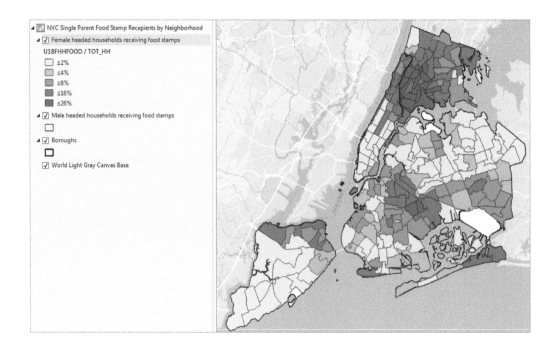

Import symbology and use swipe to compare features

To compare two maps, especially normalized segments of the same total population, you will use the same numerical and color scales for both maps. ArcGIS Pro allows you to import symbology. Next, you'll import and reuse the symbology of the female-headed households for the male-headed households. You will see many fewer male-headed households receiving food stamps.

1.   Open the Symbology pane for Male-Headed Households Receiving Food Stamps, click the Options button and click Import symbology.

2.   In the Apply Symbology From Layer pane, make the selections as shown:

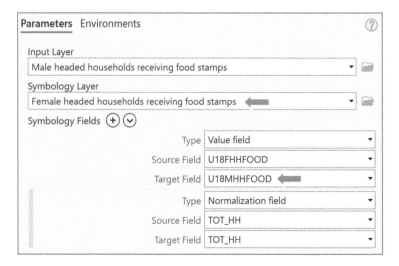

3.   **Click Run.**

4.   **In the Contents pane, click Female Headed Households Receiving Food Stamps.**

5.   **On the Appearance tab under Compare, click Swipe, and drag the pointer vertically or horizon-tally to reveal the layer underneath.** This step allows you to see the values of the Male-Headed Household Receiving Food Stamps layer without having to turn off the layer above it. You can see the same spatial patterns in both maps but with much lower percentages for male-headed households receiving food stamps.

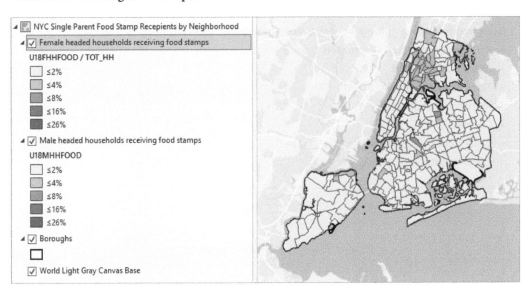

6.   **On the Map tab, click the Explore button to deactivate swipe.**

7.   **Save your project.**

Tutorial 2-7: Density maps

Density maps divide populations and other variables by their polygon areas, yielding a measure of spatial concentration. If you divide a population by its polygon areas, the resulting population density (for example, persons per square mile) can provide information related to congestion or how people are distributed across an area.

   A neighborhood with a high density of households receiving food stamps but a low density of food banks and soup kitchens may help determine potential locations for new food banks or kitchens.

Open the Tutorial 2-7 project

1.   **Open Tutorial2-7.aprx from Chapter2\Tutorials, and save the project as** Tutorial2-7YourName.aprx. The map opens with layers added but not yet classified.

2.   **Zoom to full extent.**

Create the density map

1.   **Using the following guidelines, symbolize Food bank/soup kitchens (SQ MI):**
     - **Primary Symbology: Graduated Colors**
     - **Field: FOOD_FACIL**
     - **Normalization: AREA_SQMI**
     - **Method: Manual Interval**
     - **Classes:** 5
     - **Color scheme: Blue-Purple (five classes)**
     - **Break points:** 2, 4, 8, 16, 29

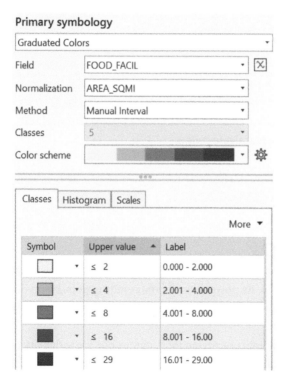

2. You can readily see high densities of food banks and soup kitchens.

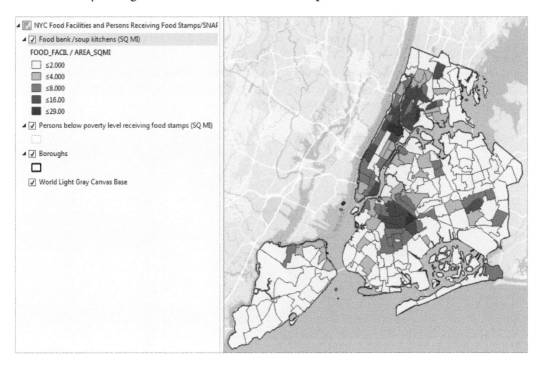

YOUR TURN

Create a density map for "Persons below poverty level receiving food stamps (SQ MI)" using Graduated Symbols. Set POV_FOOD normalized by Area_SQMI. Set **four** classes; Manual Interval; upper values **300**, **900**, **2700**, and **13000**; circle 3 with Solar Yellow color; and a symbol size range from **4** to **12**. Zoom and pan the map. Do you see any gaps with high population densities but low food facilities densities? Save your project.

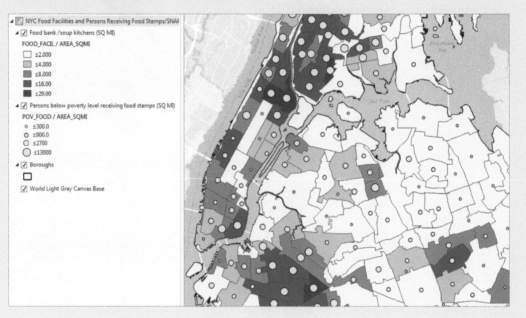

Tutorial 2-8: Group layers and layer packages

You can place layers together into groups to manage them more easily. For example, you can change the visibility of an entire group layer with one click. You can also save a group layer (or any single layer) as a layer package (a file with an.lpkx extension). A layer package is one file with all data sources and symbology included. You can share this package, including in ArcGIS Online. In this tutorial, you will create group layers for populations and facilities in New York City by administrative and political features, including fire companies and police precincts. You will then create layer packages from layer groups.

Open the Tutorial 2-8 project

1. **Open Tutorial2-8.aprx from Chapter2\Tutorials, and save the project as** Tutorial2-8YourName.aprx. The map opens with the Boroughs, Police Stations, Police Precinct Population Per Sq Mile, Fire Houses, and Fire Company Population Per Sq Mile layers already symbolized.

2.   Zoom to full extent.

Create police group layer

There are two ways to create group layers. The first is to create an empty group layer and add layers to the group layer. The second is to select existing layers in the Contents pane and create a group layer of the selection. You will use the first method to create a group layer for police.

1.   In the Contents pane, right-click the NYC Facilities and Population map, and click New Group Layer. A new layer, called New Group Layer, is created. Next, you will rename the layer.

2.   Right-click New Group Layer, and click Properties.

3.   In the Layer Properties window, on the General tab, for Name, type NYC Police.

4.   On the Metadata tab, fill in the following information:
     - **Title:** NYC Police
     - **Tags:** NYC, Police Stations, Precincts, Population
     - **Summary:** NYC Police Stations and Precinct Populations
     - **Description:** Same as Summary

5.   Click OK.

6.   In the Contents pane, drag the Boroughs layer on top of the new NYC Police layer group and release.

7.   Expand the layer group, and drag Police Stations and Police Precinct Population Per Sq Mile below Boroughs. With NYC Police expanded, you can see the group layer's member layers indented below the layer group.

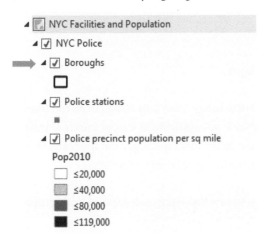

Create fire group layer

Here, you use a shortcut to create a group layer using selected layers.

1. In the Contents pane, press and hold the Shift key, and select the Fire Houses and Fire Company Population Per Sq Mile layers.

2. Right-click either layer, and click Group.

3. In the Layer Properties window, rename the new group NYC Fire.

4. On the Metadata tab, fill in the following information and click OK:
    - **Title:** NYC Fire
    - **Tags:** NYC, Fire Houses, Companies, Population
    - **Summary:** NYC Fire Houses and Company Populations
    - **Description:** Same as Summary

5. Copy and paste the Boroughs layer from the NYC Police group layer to the NYC Fire group layer.

6. Arrange the order of the layers in the group layer as shown:

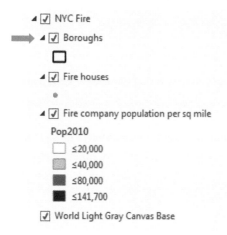

7. In the Contents pane, click the NYC Police group layer.

8. On the Appearance tab, under Effects, drag the transparency from 0 to 100%. Drag the transparency back to 0. This is an alternative to swipe that is not available in a layer group. If transparency is set to 100 percent, it will be at 100 percent in the layer package, and the features won't show up when the package is added to the new map until transparency is reduced.

9. Collapse both the NYC Police and NYC Fire groups to one line each, and expand both groups to show all layers.

Create and add a layer package

A layer package is a portable file with data and symbolization for all layers in a group or for a single layer that you can share or upload to ArcGIS Online. In addition to creating layer packages, you can create and share individual layers as layer packages.

1. In the Contents pane, right-click NYC Police, and click Sharing > Share as Layer Package.

2. Make selections or type as shown, saving the layer package file to Chapter2\Tutorials. Do not include spaces in the layer package name.

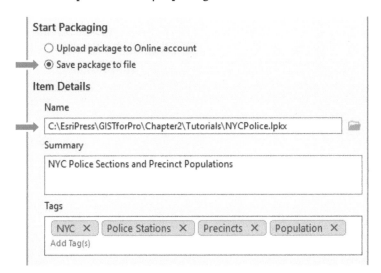

3. Click Analyze and Package, and wait for the layer package to be prepared. You will see a message stating that the layer package was successfully created. You can use your layer package in another map, share it with another ArcGIS user, or upload it to ArcGIS Online.

4. Insert a new map, and add NYC Police from Chapter2\Tutorials. The layers appear as a group and are already classified.

YOUR TURN

Create and add a layer package for NYC Fire to a new map, and save your project.

Assignments

This chapter has four assignments to complete that you can download with data from ArcGIS Online, at go.esri.com/GISTforPro2.8Data:

- **Assignment 2-1**: Analyze accessibility to charter schools in New York City.
- **Assignment 2-2**: Study K-12 population versus school type densities.
- **Assignment 2-3**: Analyze military sites for closures by congressional district.
- **Assignment 2-4**: Analyze US veteran unemployment status.

Map outputs for GIS projects

LEARNING GOALS

- Learn about alternatives for sharing maps and information from GIS projects.
- Build map layouts.
- Add visibility ranges for interactive map use.
- Build ArcGIS StoryMaps stories.
- Make professional-quality tables and charts in Microsoft Excel (optional).

Introduction

Map outputs start with the interactive maps you have been creating and using in ArcGIS Pro in chapters 1 and 2. Those maps are fine and may be all you need personally, but what about others who need spatial information—your friends, your boss or clients, a group to which you belong, stakeholders in a public policy issue, or the public at large? What if they do not have ArcGIS Pro, and if they do, what if they don't have your GIS files? This chapter focuses on providing maps that others can access, even users who may have no GIS knowledge or software. In addition to accessibility, this chapter also introduces GIS outputs, including map outputs that have nonmap elements to enable map reading and interpretation.

Consider the problem, addressed in this chapter, of college students about to get their undergraduate degrees from fine arts programs. They love their field of art, but it's difficult to find good-paying jobs in fine arts, or in some cases, any jobs at all. They're willing to relocate if they can find their dream jobs.

Suppose that a national organization for promoting fine arts education wants you to complete a GIS project—namely, to build a tool for arts students that allows them to explore employment possibilities. The tool is an interactive map that displays the number of employees and average incomes in the arts field by state and city in the United States. The tool also addresses the complex but essential issue of cost of living. For example, is a $60,000-a-year job in New York City better financially than a $30,000-a-year job in Columbus, Ohio, after cost of living is factored in? It's not, if the cost of living in New York City is more than twice that of Columbus, assuming that all income is spent on living expenses.

Suppose further, that you've already collected the data and have maps built in ArcGIS Pro. How can you get the results into the hands of arts students across the United States and around the world? In the days before ArcGIS Online and apps such as ArcGIS StoryMaps (that you'll learn to use in this chapter), there was no easy answer to this question. Now, there are good answers, with a range of old and new options available depending on your map audience's needs.

In general, to share maps from a GIS project, the map outputs must be accessible and readable, and they must also tell a story, to wit:

1. *Be accessible:* Make your maps available in hard-copy documents, via email or links, from a website or an app. Each form of output is valuable for different circumstances. A lay audience may be best served with static maps (images), printed in a document or on a presentation slide. That form of output does not require an audience to learn how to navigate an interactive map, and the author controls the desired message, with maps designed to share information. More sophisticated users may want to explore spatial information on their own, focusing on items and areas of interest to them (and unpredictable by you in advance) in an interactive app.

2. *Be readable:* Include titles, legends, scales, and other elements that allow direct interpretation of maps. These elements, called map surrounds, are essential.

3. *Tell a story:* Explain as needed the context, sources, and quality of data, problem addressed, map solution, analysis, and limitations of the solution. GIS projects include stories that have one or more maps at the center, but the projects also may need text, data tables, charts, images, and hyperlinks.

One solution is to publish maps in ArcGIS Online and make them publicly available in an app such as ArcGIS® Explorer. Then anyone can search for and interactively use your maps, or you can send anyone the URLs of the maps. That solution takes care of most accessibility and readability issues, and you already know how to publish maps online.

Another option is to create map layouts that include one or more maps and surrounds for interpretation. Then you can export a layout as an image file (for example, JPEG) and paste the file in a Microsoft Word document report or a PowerPoint slide in a presentation. You also can embed a layout in a web page to cover all three issues noted earlier, albeit with static maps. You'll learn how to create a layout in this chapter.

A third option is to use one or more maps you published in ArcGIS Online in a StoryMaps story. StoryMaps includes many apps that not only have one or more interactive maps but also allow you to include map surrounds, text, images, hyperlinks, and other materials. StoryMaps stories require little or no knowledge of website hosting, publishing, or underlying web programing.

This chapter includes two additional, supporting topics for you to use in GIS outputs. The first topic is visibility ranges, which enhance interactive map output for complex or detailed maps. The basic idea of visibility ranges is that when zoomed far out or at full extent (at small scales), you can automatically have major map layers turned on but detailed layers turned off. Then when you zoom in (to large scales), you can have detailed layers turned on when you need to see them. The second topic focuses on creating tables and charts in Microsoft Excel for use in StoryMaps stories. This tutorial is optional for those who have Microsoft Excel skills and want to make professional tables and charts and save them as screen captures to be used as resources in building StoryMaps story content.

Tutorial 3-1: Layouts

In this tutorial, you will build a map layout with two maps. Once you learn this skill, you can build any kind of layout, with one, two, or several maps, plus other elements. The purpose of the layout you will build next is to provide state-level information on the location of jobs in the arts field and average annual wages.

The layout you'll build doesn't have a layout title (such as "Maps showing arts employment, 2015"). The reason is because this layout, like most you'll produce, is destined to be used as a figure in a report, a slide in a presentation, or an image on a website. The layout title is better created in a word processer as a figure caption, in a presentation package as a slide title, or on a website.

Arts employment per 1,000 population
and annual average wages

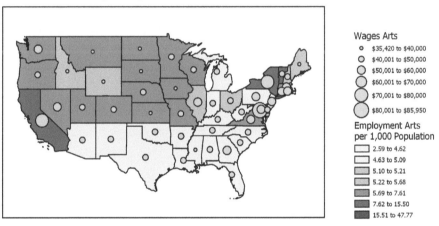

Arts employment
and annual average wages

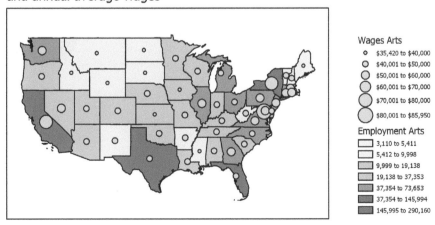

Open the Tutorial 3-1 project

Both maps for which you will need to build a layout already exist and are ready for use in the Tutorial 3-1 project file that you are about to open.

1. **Open Tutorial3-1.aprx from Chapter3\Tutorials, save it as** Tutorial3-1YourName.aprx, **and zoom to full extent.**

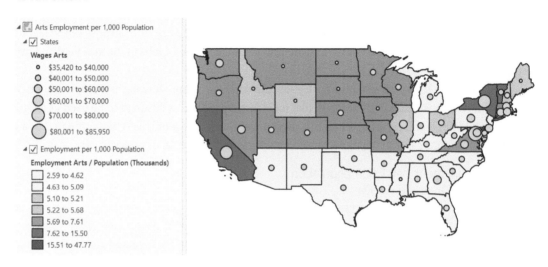

2. **Open the second map,** *Arts Employment*, **and zoom to full extent.**

Create a layout

A layout starts as a blank canvas to which you can add one or more maps, surrounds, and other elements of your choice and design. Although it's possible to add images of tables and charts to layouts, tables and charts are better placed in reports, presentations, and websites as separate tables and figures. Therefore, you can keep your layouts simple and only include maps, legends, and other common map surround elements.

1. **On the Insert tab, click the New Layout button**, **click ANSI – Portrait, and click Letter 8.5" × 11".** You can widen your ArcGIS Pro window if you don't see the New Layout button.

2. **In the Catalog pane, rename the layout** Arts Employment Layout.

Add maps to the layout

1. **In the Contents pane, right-click Arts Employment Layout, and click Zoom To Page.**

2. **On the Insert tab, in the Map Frames group, click the Map Frame arrow, select the *Arts Employment per 1,000 Population* map with the Default Extent, click and drag a bounding box in the top half of the layout, and release the mouse.** This adds the selected map to the layout.

3. **Likewise, insert the *Arts Employment* map in the bottom half of the layout.**

Resize and place the two maps

Next, you must resize the two maps in the same dimensions and set their positions. Setting the dimensions is not an exact determination, except it is recommended that the maps be the largest elements in the layout. Planning and trial and error are involved in getting a layout to look right. This exercise provides a set of dimensions and placement locations to save you time and ensure your results match those of the finished layout shown at the beginning of this tutorial. In a later exercise to add legends to the layout, you'll use an alternative, graphic method of sizing and locating layout elements.

1. **Right-click the *Arts Employment per 1,000 Population* map (blue color scheme), and click Properties.**

2. **In the Format pane, click the Placement button**, **type** 5.5 **for width and** 3.5 **for height, and type** 0.5 **in. for X and** 5.75 **in. for Y.**

3. **With the Format pane still open, click the second map (*Arts Employment*, green color scheme), type its dimensions to be the same (**5.5 **by** 3.5**), and type** 0.5 **in. for X and** 1.50 **in. for Y. Close the Format Map Frame pane.**

Next, you'll change both maps in the layout to full extent.

4. **On the Layout tab in the Map group, select each map, click the Full Extent button, and click the Fixed Zoom-In button once so that the maps fill their frames.**

5. **Save your project.**

Add guides

In this exercise, you'll use guides to position objects in a layout (legends, in this case). The guides are available from the horizontal and vertical rulers bordering the layout. When you drag objects, such as a map frame, they snap to guides. If there are intersecting vertical and horizontal guides, you can snap a corner of an object to the intersection.

1. **If your layout does not have horizontal and vertical rulers, right-click in the white area of the layout and select Rulers.**

2. **Right-click the vertical ruler at 5, and click Add Guide. Do the same at 9.25.** If you need to adjust your guide placement, hover in the ruler area, and click the arrow.

3. **Right-click the horizontal ruler at 6.25, and click Add Guide, and then do the same for 8.** You'll place legends within the guides and bottoms of maps.

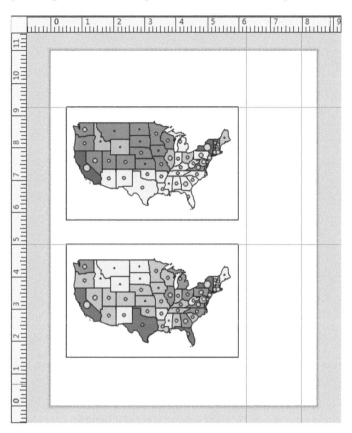

Insert legends

Using ArcGIS Pro, you'll build a legend for all layers of a map that are turned on (all layers are turned on, in this case). The legend is dynamic: if you change symbolization on the map, turn on or off layers, and so on, the legend automatically updates in the layout.

1. **Click the top map to make it active.**

2. **On the Insert tab, in the Map Surrounds group, click the Legend button, and drag a rectangle that snaps to the tall and narrow rectangle that your guides have formed on the right side of the map.** ArcGIS Pro creates and draws the legend, but it does not draw completely, as indicated by the red dots in parentheses. The labeling for the choropleth map for the legend is too wide to fit in the layout. In the next exercise, you'll fix this problem so that the entire legend is visible.

3.   **Click the bottom map in the layout, and create its legend as shown.** This legend's labels are not too wide, so the legend draws completely within the available space.

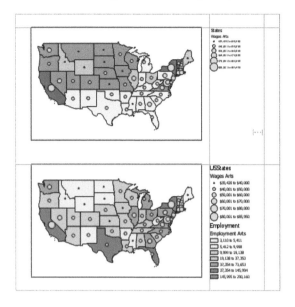

Modify legends

Next, you will convert the top legend to a static graphic, ungroup legend elements, edit them to be narrower, and regroup them.

1.   **Click the top legend, drag a selection handle on its right side to the right beyond the edge of the layout, and release the handle so that the entire legend draws.**

2.   **Right-click the top legend, and click Zoom To Selected.**

3.   **Right-click the top legend again, and click Convert To Graphics.** This action causes the legend to be static so that it no longer automatically updates if you turn layers on or off on maps or make other changes to the map.

4.   **Right-click the legend symbology, and click Ungroup.** Now the two parts of the legend are separate.

5.   **Right-click the States area of the legend and click Ungroup again.** The upper elements of the legend are now separate graphics.

6.   **Ungroup the Employment per 1,000 Population area of the legend.** All elements of the legend are now separate graphics.

7.   **In the top part of the legend, click States, and press the Delete key.**

8.   Click Employment per 1,000 Population, and press the Delete key.

9.   Double-click the Employment Arts/Population (Thousands) text box, and in the Format Text pane, click the Options button.

10.   Under Text, in the Text box, place your pointer after Employment Arts, press Shift and Enter, and edit the text of the second line to per 1,000 Population.

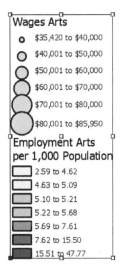

11.   Adjust the text box for Employment Arts to display both lines of the legend heading.

12.   Finally, select all elements of the legend by dragging a rectangle around them, right-click the selection, and click Group. Now the legend is back to being one graphic.

YOUR TURN

Modify the bottom legend, deleting the USStates and Employment text boxes. Move the top half of the legend down a bit to eliminate the extra space between the two halves. Regroup the legend. Zoom to Page when you finish. Save your project.

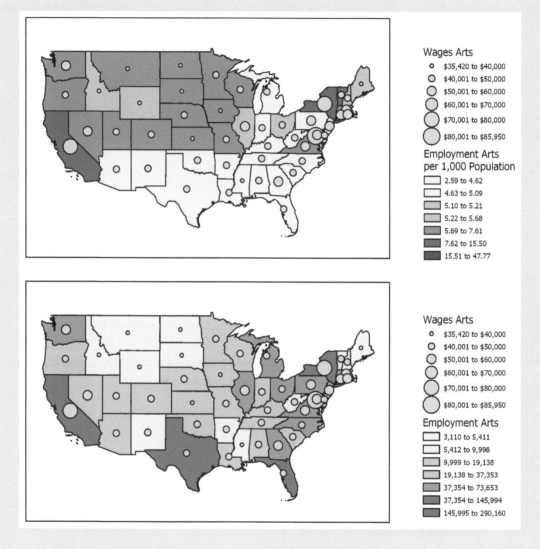

Insert text

Next, you'll insert text, titling each map (but not the whole layout).

1. On the Insert tab, in the Graphics and Text group, click the Straight Text button **A**.

2. Click above the upper-left corner of the top map.

3.   **Double-click the new text element to open the Format Text pane.**

4.   **Under Text, click the text box, and type** Arts employment per 1,000 population, **press Shift and Enter, and type** and annual average wages.

5.   **In the Format Text pane, click Text Symbol > General > Appearance. Change the size to** 16 **pt, and click Apply.**

6.   **Move the text above the map and left align it.**

Arts employment per 1,000 population
and annual average wages

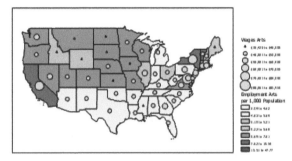

YOUR TURN

Insert text for the bottom map. Type **Arts employment**, press Shift and Enter, and type **and annual average wages**. Change the font to **16** pt. To finish, press the Shift key, select both text boxes and the two maps (but not the legends), right-click the selection, click Align, click Align Left, and save your project.

Tutorial 3-2: Visibility ranges

This tutorial shows you how to set visibility ranges for map layers and feature labels. GIS uses visibility ranges to automatically turn map layers on and off, depending on how far you're zoomed in or out of your map. Before you work with visibility ranges, you will learn about map scale, the underlying measure that controls visibility ranges.

Suppose that you have two points on your map, A and B, which are 1 inch apart on your computer screen. Map scale is the ratio of the distance between A and B (1 inch) on your computer screen divided by the distance between the same two points in inches on the ground. An example scale for displaying the Lower 48 states is 1:50,000,000, meaning that 1 inch on your computer's map corresponds to 50,000,000 inches on the ground. Map scale is unitless, as a ratio that divides units, so, for example, 1:50,000,000 can also mean 1 foot on your map equals 50,000,000 feet on the ground. You can use any distance unit. The map terms "small scale" and "large scale" are counterintuitive, so understanding the next three sentences is important. The scale 1:50,000,000 is called small scale because the ratio 1/50,000,000 is small, even though the map shows a large area. And the scale 1:24,000 is considered "large scale," because its ratio is relatively large, even though the map shows a small area. All map scales are considerably less than 1 because 1:1 is life size, and all maps represent land areas much larger than the maps themselves.

Next, you'll set visibility ranges to view arts employment at the state level when the map opens at full extent, so when you zoom in close enough, employment at the metropolitan (city) level will turn on. First, you will work on arts employment across the United States. Then you can zoom in to a state of interest to get employment in a metropolitan area.

Open the Tutorial 3-2 project

1. **Open Tutorial3-2.aprx.** The project opens with all layers turned on and in a jumble. You'll set a visibility scale to automatically turn off the metropolitan points at the full map extent.

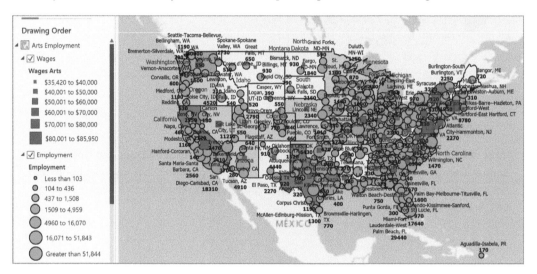

2. **Save the project as** Tutorial3-2YourName.aprx **in the Tutorials folder.**

Set visibility ranges for map layers

Map scale depends on the size of your computer screen and the size of your map view (the larger your map view, the larger the map scale), so the steps that follow cannot use specific map scales for triggering visibility ranges. Instead, the project has spatial bookmarks you'll use to get appropriate map scales for your specific screen.

1. **Use the Arizona bookmark.** At this map scale (1:8,324,191 in the figure that follows but different on your computer) and further zoomed in, you'll set visibility ranges to display the Employment point layer and turn off the display of Wages. You can turn off Wages when you zoom in, because

if you were the map user, you'd already know Arizona's average wages in the arts industry from the map of the 48 states, which is the starting view.

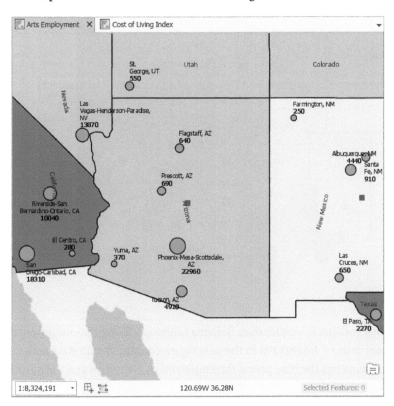

2. In the Contents pane, right-click the Employment point layer (symbolized in blue) of the *Arts Employment* choropleth map and click Properties.

3. On the General tab, in the Visibility Range section, click the Out Beyond (Minimum Scale) arrow, select <Current>, and click OK. So, when you're zoomed out beyond the current scale of the Arizona bookmark and up to full extent, the Employment point layer will be turned off. If you zoom in closer than the Arizona bookmark, the layer will turn on.

4. In the Contents pane, right-click the Wages layer, and click Properties.

5. On the General tab, in the Visibility range section, click the In Beyond (Maximum Scale) arrow, select <Current>, and click OK. Now Wages will display when zoomed out farther than the Arizona bookmark and turn off when zoomed closer in than the bookmark. Next, you'll test the two visibility ranges.

6. **Zoom to full extent.** Employment Arts has no visibility range and is always on. Wages is on because you're zoomed out from Arizona, but Employment is turned off for the same reason.

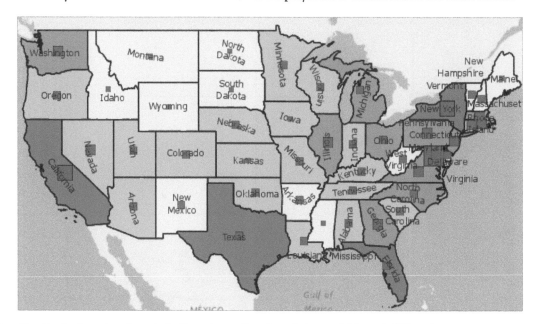

7. **Use the Indiana bookmark.** Indiana is smaller than Arizona (although filling the computer screen, Indiana has the larger map scale of 1:6,880,105 in the next figure compared with Arizona's 1:8,324,191); therefore, Indiana has the blue points for Employment turned on and the purple square points for Wages turned off.

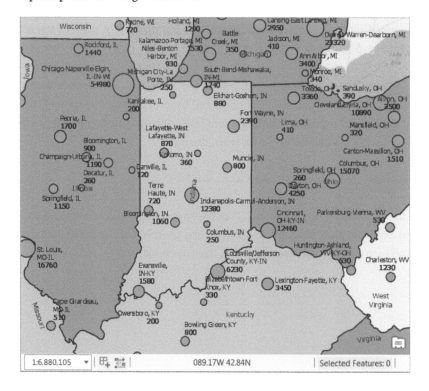

8. **Zoom to full extent, and save your project.**

Set a visibility range for labels

You can turn a feature layer's labels on and off using visibility ranges, too. For example, when at full extent, it's helpful to have state names displayed as labels, but when zoomed in, the labels are no longer needed and amount to a small amount of clutter.

1. **Use the Arizona bookmark.** The objective is to turn off the state name labels when zoomed in closer than Arizona.

2. **In the Contents pane, select the Employment polygon layer (symbolized in green).** The Feature Layer contextual tab for this layer appears at the top of the ArcGIS Pro application.

3. **Click the Labeling tab.** This label already has labeling on and configured but so far does not have a visibility range.

4. **For In Beyond in the Visibility Range group of the Labeling tab, select <Current>.**

5. **Zoom to full extent.** The state labels are on.

6. **Zoom to Indiana.** The state labels now turn off at this scale, while the labels for the cities (from the Employment point layer) remain on.

7. **Save your project.**

YOUR TURN

Open the *Cost of Living Index* map. For now, it's sufficient to say that certain West and East Coast places are a lot more expensive to live in than other parts of the country, as seen on the map that follows. Here, your tasks are to set visibility ranges like those you just did. Set the visibility range for the Cost of Living Index point layer so that the layer turns on when zoomed in to the Arizona bookmark and beyond. Turn off Wages and the labels from the States layer when zoomed in beyond the Arizona bookmark. Save your project when finished.

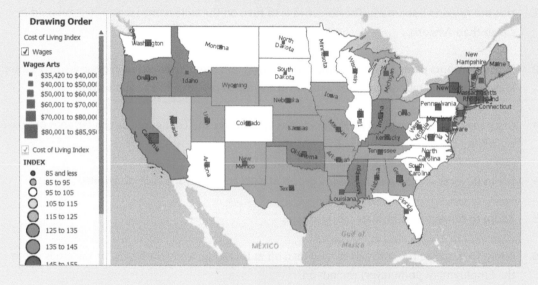

Tutorial 3-3: ArcGIS StoryMaps: Preparing the data

The StoryMaps app allows you to include web-based interactive maps, map surrounds, paragraphs of text with extensive formatting, images, videos, hyperlinks, and other content. The maps in StoryMaps stories are fully interactive, including the usual navigation, plus pop-ups and visibility ranges that you create in ArcGIS Pro.

You will use the following plan for building a story on arts employment:

1. You'll publish two of the three arts employment maps that you've been using in this chapter to your ArcGIS Online account. You'll leave out the *Employment per 1,000 Population* map, which does not play a major role.

2. You'll use the StoryMaps app for presenting the two maps on the web, each fixed in place with scrolling text on the left of each map. This layout allows a map to be the focus for viewing and interaction, while text and other content describing the maps are available for use as well.

3. With the maps in place, to save time, you'll copy and paste text and images (for telling the story) that are already written and created, and available in Chapter3\Tutorials\Resources. Although a convenience in this tutorial, it's good practice to create a manuscript in Word or other word processing software as well as all exhibits before you start building a website such

as a story. Then you can combine the work of coauthors, have content reviewed and accepted, and get organized without dealing with the technicalities of the StoryMaps app at the same time.

In general, you will use the following workflow to build a story:

- You'll author your maps in ArcGIS Pro, provide aliases for attributes, add and symbolize map layers, and create visibility ranges.
- You'll share, or publish, your maps to your ArcGIS Online account, where you configure pop-ups and do some labeling.
- You'll sign in to storymaps.arcgis.com in your browser to view your Stories page where you can create a story as well as access all your stories after you create additional ones.

The next guidelines will help you successfully publish your ArcGIS Pro maps to ArcGIS Online. These guidelines use material from chapters 4 and 5, so you can refer to this paragraph after you finish those chapters and whenever you create stories.

- Keep ObjectID and Shape fields visible in all layers (open Fields View and ensure those fields are checked). See chapter 4.
- Make your joins of tables to maps permanent (right-click the layer with a join, click Data > Export Features, and use the result for mapping). See chapter 4.
- Although not essential, it's a good idea to give your maps and all feature classes the WGS84 Web Mercator (Auxiliary Sphere) projection. This projection will guarantee that your map layers display online and will publish your map faster online. Use the Project tool to change projection of a map. See chapter 5.
- Also, although not essential, you can make a new map layer using the Feature To Point tool to map center points of polygons if you are symbolizing center points of polygons (for example, with graduated size point symbols for a quantitative attribute). This process will speed up rendering your map online. Check the Inside option to make all centroids lie inside their polygons. See chapter 4.

Open the Tutorial 3-3 project

1. **Open Tutorial3-3.aprx.** This tutorial has two of the maps from the finished tutorial 3-2, with visibility ranges added.

2. **Save the project as** Tutorial3-3YourName.aprx **to the Tutorials folder.**

3. **Right-click each layer in the Contents pane, and click Disable Pop-ups for all map layers of the** *Arts Employment* **map.** You'll configure pop-ups in ArcGIS Online after the map is published.

4. **Similarly, click Disable Pop-ups for the** *Cost of Living Index* **map.**

5. **Save your project.**

Create label fields for use online

Your maps for metropolitan areas need labels that combine data from more than one field. Next is an example pop-up as needed for the Metropolitan Employment layer of the *Arts Employment* map:

> Cleveland-Elyria, OH, Employment = 10890, Wages = $46150

Here, Cleveland-Elyria, OH is from the Name field, 10890 is from the employment field, and $46150 is from the Wages field of Metropolitan Employment. All characters not mentioned in the previous sentence are hardcoded text that is constant. The Python scripting language expression that creates the label value, stored in a field named Label, is as follows:

```
!NAME! + ", Employment = " + str(int(!Employment!)) + ", Wages = $" +
str(int(!Wages!))
```

The + sign is the concatenation operator that concatenates (or adds together) text from different sources to create a text value. Any string of text enclosed in exclamation points, such as !NAME!, retrieves values from a field—the Name field, in this case. Any string of text enclosed in double quotation marks is constant. It's only possible to concatenate text values, but Employment and Wages are floating point numbers. The int() function converts floating point values into integers (to eliminate decimal places), and the str() function converts numeric values into text. You'll label layers in ArcGIS Online, but it's convenient to have data for a label already assembled in a field, so you'll calculate label fields in ArcGIS Pro before publishing web layers.

1. **Open the attribute table for Metropolitan Employment.** This table already has a text attribute named LABEL with the correct labels computed, and a second text attribute named LabelPractice, in which you will re-create the labels for practice.

2. **In the table, right-click the LabelPractice heading, and click Calculate Field.** You may have to wait a minute while ArcGIS Pro creates a cache for geoprocessing.

3. **Create the expression as follows: !NAME! + ", Employment = " + str(int(!Employment!)) + ",** **Wages = $" + str(int(!Wages!)).** You can add fields by double-clicking the Fields panel and clicking the + from the list of operators, but you must type everything else including spaces.

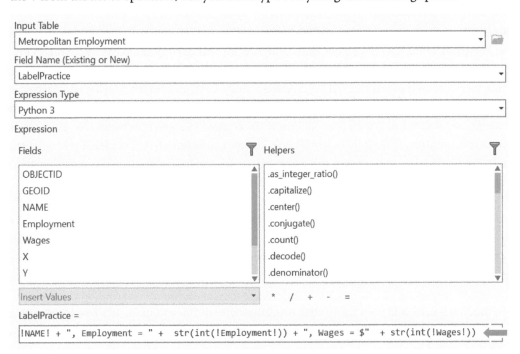

4. **Run the Calculate Field tool.** If you get an error message, review your expression carefully. If needed, you will correct the errors and run it again. You will see values such as those shown in the LABEL field of the attribute table.

YOUR TURN

Open the *Cost of Living Index* map, and in the Metropolitan Cost of Living Index layer, calculate the following values for the LabelPractice field:

```
!NAME! + ", Index = " + str(int(!INDEX_!))
```

Label already has the values from the expression computed, but recomputing them in LabelPractice is good practice.

Share maps in ArcGIS Online

You must first share (publish) maps in ArcGIS Online for use in the StoryMaps app and other ArcGIS Online apps. You have two maps to share: *Arts Employment* and *Cost of Living Index.*

1. **In File Explorer, browse to Chapter3\Tutorials\Resources, and open StoryMapManuscript.docx in Microsoft Word.** The document starts with text for you to copy and paste into ArcGIS Pro when you share the employment maps. Additional text is for the content that accompanies maps that you'll use in the next tutorial for creating a story. Each section for copying and pasting is numbered for easy reference. Note that it is good practice to compose content in a nonweb document such as Word, and then copy and paste the content to apps or websites. For example, if you have collaborators, it's helpful to use Word's Review capabilities to collect feedback, corrections, additions, and so on. Then when you create the corresponding website, you can concentrate on technical issues and be confident that the content is stable and ready for use.

2. **With the *Arts Employment* map open, on the Share tab, in the Share As group, click Web Map.**

3. **Zoom your map to full extent.** This step is necessary so that the default extent for the map in ArcGIS Online is full extent.

4. In the Name field, **add your name after Arts Employment (for example, Arts Employment** Your Name). All maps stored in an organizational account must have unique names. Also, if your work is being graded in a class, adding your name allows an instructor to identify your work.

5. In the Summary field, **copy and paste the Summary 1-1 text from the StoryMapManuscript document.**

6. **In the Tags field, copy and paste tags 1-2.**

7. **Under Share With, select Everyone.** You must share your maps used in a story with everyone; otherwise, you will be the only one able to view your story. You can also share only with your organization or a specified group from your organization, instead of the general public.

8. **Click Analyze.** No problems are anticipated in uploading the map.

9. **Click Share.** Publishing the map may take a few minutes.

YOUR TURN

Similarly, share the *Cost of Living Index* map, copying and pasting text from Summary 2-1 and Tags 2-2 of the StoryMapManuscript.docx. Remember to click Share > Web Map, zoom to full extent, and fill in the Share As Web Map pane.

Move online maps to a new folder

You'll publish more maps in future tutorials and assignments, so to avoid clutter in My Content in ArcGIS Online, you'll create a folder for all the maps and resources of the *Arts Employment* story. Then you'll move the maps published online to the new folder. Note that even though you'll have files in different folders, all files stored in your organizational account must have unique names (including files of other users in the same organizational account). In contrast, files in different folders in your Windows file system can have the same names.

1. **Open ArcGIS Online** (arcgis.com) **in a browser, and sign in.**

2. **Go to Content, and under My Content, click the Create New Folder button** ⊟⁺. **Type** ArtsEmploymentStoryMapYourName **for the folder name, click OK, and click the All My Content button.**

3. **Select the check boxes on the left of the three Cost of Living Index Your Name and three Arts Employment Your Name files in the item list.**

4. **Click the Move button** ⟨↕⟩.

5. **Click the ArtsEmploymentStoryMapYourName folder > Save.**

6. **In the Folders list, click the ArtsEmploymentStoryMapYourName folder.** Your six files are there.

Add labeling

Next, you'll use one of the Label fields created in ArcGIS Pro to label a map layer.

1. **Open the** *Arts Employment Your Name* **web map in Map Viewer.**

2.   Click the Default Map View button ⌂.

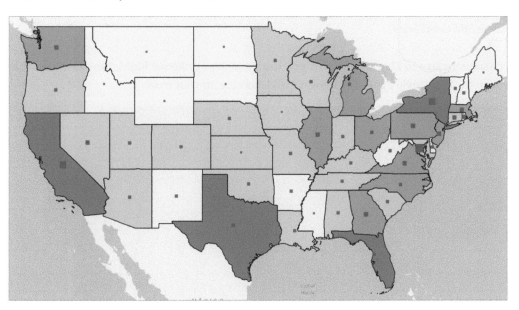

3.   Zoom in and pan to Southern California so that the Metropolitan Employment point layer becomes visible.

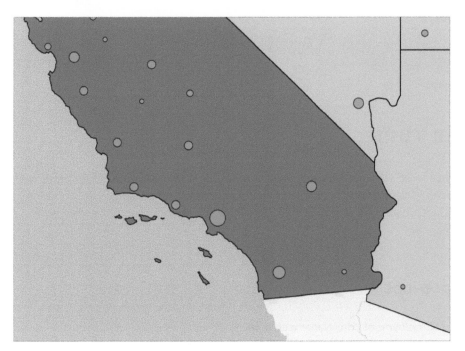

4.   In the Layers pane, click Metropolitan Employment. Then from the menu on the right, click Labels.

5.   Click Add Label Class, under Label field, click NAME.

6.   **Click LABEL and Replace.**

7.   **Click Edit Label Style.**

8.   **Change the size to 11, the color to black, the placement to Above Right, turn the halo off, and** close the Label style pane. The reformatted labels are on your map.

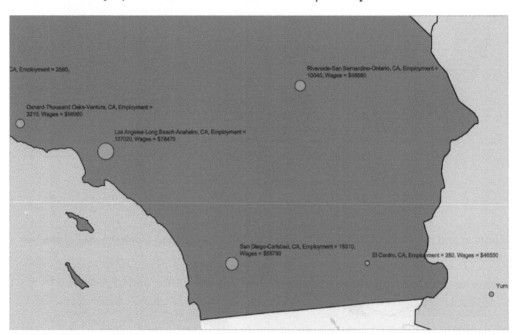

9.   **Save your map.**

YOUR TURN

Go to Content, and in your ArtsEmploymentStoryMapYourName folder, open the *Cost of Living Index Your Name* map. With the map at Default Map View, label the State Cost of Living layer with Name and a size 11 black font. With the map zoomed in so that the Metropolitan Cost of Living Index layer is visible, label that layer with the Label field, a size 11 black font, and a white halo. Save your map.

Configure pop-ups

1.   **Open the *Arts Employment Your Name* web map.**

2.   **In the Layers pane, click State Employment, and from the menu on the right, click Configure Pop-Ups.**

3.   **In the Configure Pop-Ups pane, enable pop-ups.**

4. **Click Fields to expand the list, and click Select fields.**

5. **Click Select All and then Deselect All.**

6. **Click attributes so that only the following are checked: Name, Employment Arts, Population, Wages Arts.**

7. **Click Done.**

8. **Zoom out so that you can see several states, and click any state to try out the pop-up.**

9. **Save the map.**

> ## YOUR TURN
>
> Configure pop-ups for the *Cost of Living Index Your Name* map. For the State Cost of Living Index layer, use Name, EmpArts, Index, and Wage Arts. Save your map, and sign out of ArcGIS Online.

Tutorial 3-4: ArcGIS StoryMaps: Creating a story

The online maps for the arts employment story are ready to use, so next you'll create a story, add the maps to it, and add nonmap content.

Create a story and its cover

Text and images of charts and tables are in the Chapter3\Tutorials\Resources folder, ready for you to copy and paste or add to your story.

1. **In File Explorer, go to Chapter3\Tutorials\Resources and open StoryMapManuscript.docx.** It will be convenient for you to leave the Resources folder open so that later in this tutorial you can drag image files from it to the StoryMaps app rather than browsing for them.

2. **Open a browser, go to storymaps.arcgis.com, and sign in.** That action puts you on your own stories page. Of course, you don't have any stories yet. After you create stories, they each appear on this page as cards, with cover page image, date created, title, and subtitle.

3. **Click the Start a Story button.**

4. **From StoryMapManuscript.docx, copy and paste Title 3-1 and Subtitle 3-2 into the StoryMaps cover page.** Notice in the top horizontal menu that whenever you add new content or make

changes, the StoryMaps app automatically saves your work. Currently your story is classified as a draft because you have not yet published it for use on the web.

5. In your browser, search for free arts images and select an image of your choice, with width longer than height, for the cover page, saving it to Chapter3\Tutorials\Resources. An example website is pexels.com, which does not require credits for authorship. Nevertheless, save the URL where you get your image so that you can give credit in the Credits section of your story later. Alternatively, you may use the image used in this example, pexels-engin-akyurt-6137963.jpg, downloaded from www.pexels.com/photo/light-red-art-blue-6137963/. It is in the Resources folder for your use.

6. In StoryMaps, click the Add cover image or video button, and add the image of your choice.

7. At the top of the window, click the Design button, and under Cover, choose Full.

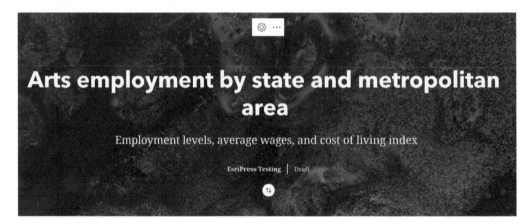

8. Click the Options button ⚙, and under Properties, paste Alternative Text 3-3. This text is available for people with impaired vision. Search engines also can access this text.

Add blocks for the introduction

Each separate element of content in a story is called a block. Examples of blocks are headings, paragraphs, bulleted lists, images, maps, and so on. Each time you need to add content of a different type, you will open the block palette, chose the needed block type, and add the content. Generally you will add blocks in the order of your story, but you can always insert new blocks where needed or move blocks up or down.

1. Scroll down if necessary, and click the Tell Your Story button. The block palette opens.

2. Click the Text button. From the menu, select Paragraph and change the selection to Heading. Copy and paste Heading 4-1. Later in this tutorial, you will add navigational links to your cover page. The links are the heading texts, so it is recommended that you design headings to be meaningful and consistent (for example, All Subject Titles).

3. Hover below and to the left of the heading, click the Add Content Block button that appears, click Text, select Paragraph and change the selection to Subheading. Copy and paste Subheading 4-2.

4. Add a Paragraph block, and copy and paste Paragraphs 4-3.

5. Add a Subheading block, and copy and paste Subheading 4-4.

6. Add a Paragraph block, and copy and paste Paragraph 4-5.

7. In the first paragraph block, select the last sentence of the second paragraph starting with "You might be better off…" A rich text editor appears.

8. In the editor, click the italic button. The sentence is now in italic.

Add a sidecar block for the employment map

A sidecar block with docked panel has media (your interactive employment map) in a panel on the right and text and other content (for example, charts or tables) in a scrolling panel on the left of a vertically split screen.

The docked panel is useful when you have a lot of narrative to enter in the left panel, as is the case of the employment map. The floating panel option, shown as an option below, is for cases with short narrative (a caption or a few sentences), and maximizes the size of the media (in this case, a map).

1. Click the Add Content Block button, select Sidecar, and click Docked Panel.

2. Click the Add Media drop list, click Add Map, and click Arts Employment Your Name. Your maps appear in cards, most recent first, with a thumbnail of map, date edited, map name, and summary. Next is an interface allowing you to configure your map display.

3. Click the Zoom In button and pan by dragging the map until the Lower 48 states are centered and fill the map panel.

4. Click the Bookmarks button. Two bookmarks are available, Arizona and Indiana.

5. Click the Options button and turn on Search and Legend. Buttons for those options appear on the map.

6. In the map, click the Legend button. The legend appears for the visible layers.

7. Click the Place Map button. The configured map appears in your story.

8. Click the Options button at the top of the map, copy and paste Alternative text 4-6, and click Save.

Add content for the left panel of the employment map

1. In the left panel, click the Continue Your Story button, click Text, and select Heading. Copy and paste Heading 5-1.

2. Add a Subheading block, and copy and paste Subheading 5-2.

3. Add a Bulleted list, and copy and paste Bulleted List 5-3. Place the pointer at the start of each sentence, and press Enter to create bulleted items.

4. Add a subheading, and copy and paste Subheading 5-4.

5. Add a paragraph, and copy and paste Paragraph 5-5.

6. Add an image, drag ArtsWagesChart.jpg from the Resources folder, and below the image, copy and paste Caption 5-6.

7. Add an image, drag Table2, and add Caption 5-7.

Preview and publish your map

Next you will preview your map as it would appear on a smartphone, tablet, or your computer monitor and publish it to the internet.

1. On the top horizontal menu, click Preview.

2. Click the Preview On Desktop button 🖵.

3. Try the other two previews, On Phone and On Tablet.

4. Click Edit Story.

5. Click Publish > Everyone (Public), and then click Publish Story.

6. At the top of the page, click the link button to copy the story URL.

7. Open a new tab on your browser, and paste the URL to open your story. Your story, although not yet finished, is on the internet as expected.

8. Close the tab, and click the Edit Story button.

YOUR TURN

Now you will add a second docked sidecar. It will have your Cost of Living map as the media and text and charts from the manuscript document and Resources folder. Start by scrolling down at the bottom of your story, and click Continue Your Story. Follow along in the manuscript: start with Heading 6-1 and work through Paragraph 7-2.

  If you want to delete a heading, subheading, or paragraph, select the text and press the Delete key. Other blocks have trash can icons for deletion, either always visible or when you hover over them. If you want to insert a new block, hover between blocks and the Add Content button appears for you to create new content. If you want to move a block, hover over it and drag the six dots that appear on its upper left to a new location.

Add navigation

Next, you will add heading links to corresponding parts of your story.

1. On the top horizontal menu, click Design.

2. Enable Navigation and Credits, and then close the panel. Notice that the headings from your story are now links in a horizontal navigation bar in your story.

3. Try out the links.

Add credits and publish your story

As is the case for anything you author, you must add credits to anything you create that uses the work of others.

1. Scroll to the bottom of your story. The Credits section appears.

2. Type Credits for the title.

3. Using the Credits part of the manuscript, copy and paste Content (first part of each entry) and Attribution (second part of each entry including URL).

4. Click the Publish button. Try out your story. You can send the story's URL, as copied earlier from ArcGIS Online or as seen at the top of your browser, for anyone to use.

Tutorial 3-4: Making charts for feature attribute data in Microsoft Excel *(Optional)*

Microsoft Excel works well with StoryMaps by creating charts and tables to show project results. If you need these skills, the next tutorial will help you create professional-quality tables and charts. You'll learn how to export attribute data from a map layer in ArcGIS Pro; import the data in Excel; clean up and sort the data; make bar charts and scatter plots; and use a Windows app to save the charts as images for insertion into a Word document, PowerPoint slides, a website, or stories in StoryMaps.

Open the Tutorial 3-4 project

1. Open Tutorial3-4.aprx in Chapter3\Tutorials, and save it as Tutorial3-4YourName.aprx.

2. Use the Lower 48 States bookmark.

Limit the visibility of attributes

Only three fields from the USStates attribute table are needed for chart making. By making only those fields visible, only they will be exported along with some additional fields that ArcGIS Pro considers essential.

1. Open the USStates attribute table.

2. Click the Options button at the upper right of the attribute table, and click Fields View.

3. At the upper left of the Fields View table, deselect Visible so that no fields are selected.

4. Select the NAME, WageArts, and Index fields to be visible.

5. Save the changes, and close the Fields table, but leave the attribute table open. Now the attribute table shows only the three needed fields.

Export data

1. Click the Options button (upper-right corner of table), and click Export.

2. Click the Browse button of the Output Location, and browse to Chapter3\Tutorials.

3. For Output Name, type USStates.txt, and click OK. The table is exported.

4. Save and exit ArcGIS Pro.

Import data to Microsoft Excel

Next, you'll import USStates.txt to Excel.

1. Start Microsoft Excel with a blank workbook.

2. Click File > Open, and browse to Chapter3/Tutorials.

3. In the Open window, on the right of File name at the bottom, select All Files.

4. Click on USStates.txt (not USStates.txt.xml), and click Open.

5. In the Text Import wizard, confirm that the Delimited Option button is selected, and click Next.

6. Select the Comma check box, and click Finish.

7. Click File > Save As, change the Save As type to Excel Workbook, and click Save.

Delete unneeded columns

1. Click the *A* column heading cell to select that column, right-click the selection, and click Delete.

2.   **Likewise, delete the Shape_Length and Shape_Area columns.**

	A	B	C
1	NAME	WageArts	Index
2	Alabama	41750	90.71111111
3	Arkansas	41070	91.325
4	California	70440	138.1666667
5	Connecticut	56460	134.7
6	Florida	47930	101.19

Sort the table

Creating a bar chart from raw data requires manual work, but it is not difficult. The preparation process requires sorting data by the column being used to make the chart, taking notes, counting frequencies, and typing information.

1.   **Select columns *A*, *B*, and *C*.** You must select all columns of a table when you sort; otherwise, you'll destroy the data.

2.   **Click Data, and click Sort.**

3.   **Ensure that you check the My Data Has Headers check box, select WageArts as the Sort By column, and click OK.** The top row of data is South Dakota with WageArts 35420 and Index 101.3.

4.   **Save the workbook in the Chapter3\Tutorials folder, and close the workbook.**

Create a bar chart

To get data for a bar chart of wages of arts employees by state, you would have to count the number of raw data records in ranges for the following wages: $40,000 or less, $40,000–$50,000, $50,000–$60,000, …, $80,000 and up. The sorted data of the previous exercise makes this task easy even when done manually, but you won't have to do the counting because an Excel workbook in the Resources folder already has the counted data. By convention, the left side of each class (for example, the $35,000 in $35,000–$40,000) is included in the class if there are any states with that value, but the right side ($40,000) is not. The $40,000 is in the next class, $40,000–$45,000.

1.   **Open the ArtsWagesChartTable.xlsx workbook from the Resources folder.**

$40K and Less	10
$40K to $50K	26
$50K to $60K	13
$70K to $80K	2
$80K and Up	1

2. To select all the data in the Chart sheet, click in cell A1, press the Shift key, and click in cell B5.

3. Click the Insert tab. In the Charts group, click the button for Insert Column or Bar Chart. Select the first 2D Column chart, Clustered Column. This step creates the chart as shown.

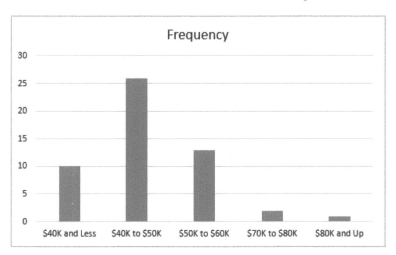

4. Save and close the workbook.

Create a scatter plot

When you create a scatter plot in Excel from two columns of data, the first column appears on the horizontal axis and the second on the vertical. In this case, to place the Cost of Living Index on the horizontal axis, you'll move the Index attribute to the left of the WageArts attribute.

1. Open the USStates.txt.xlsx workbook from Chapter3\Tutorials.

2. Select column C (Index), hold down Shift and move your pointer to the edge of the column. The four-headed arrow appears.

3. Click and drag your pointer to be between the A and B columns, and release the mouse and Shift key. Now the columns are ordered Name, Index, and WageArts.

4. Select data in the range B2 to C50.

5.   **Click the Insert tab. In the Charts group, click the Insert Scatter or Bubble Chart button, and click Scatter.** This step produces the following chart, which needs some refinements.

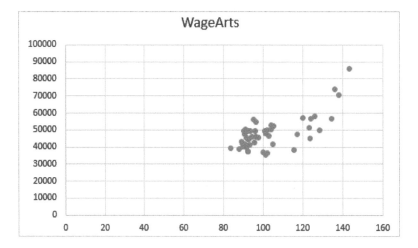

Modify the scatter plot axis ranges

Next, you will limit the ranges of the vertical axis and horizontal axis so that you can see the pattern more clearly.

1.   **Right-click any vertical axis number (such as 40000), and click Format Axis.**

2.   **Change Bounds Minimum to** 35000 **and Bounds Maximum to** 90000.

3.   **Right-click any horizontal axis number, and click Format Axis.**

4.   **Change Bounds Minimum to** 80 **and Bounds Maximum to** 150.

5.   **Adjust the size of the chart so that it's about square.**

Format an axis

1.   **Right-click any vertical axis number, and click Format Axis.**

2.   **In the Format Axis panel, click Number (you may have to scroll down). For Category, select Currency, type** 0 **for Decimal Places, and close the Format Axis panel.**

Label each axis

Your user interface may vary from the one used in this exercise, depending on your Microsoft Excel version.

1. Click anywhere in the chart.

2. From the contextual menu, click the Chart Elements button ➕, select Axis Titles, and deselect Chart Title.

3. Edit the axis labels so that the horizontal axis is Cost of Living Index and the vertical axis label is Average Arts Income.

Modify the point symbol

1. Click anywhere in the chart, and under in the Design tab, click Change Colors, and click Monochromatic Palette 7. This step completes the chart, which is now informative as well as professional in appearance. You can see that wages in the arts profession increase when the cost of living increases, and even at an increasing rate for states with a higher cost of living. The chart has no title because it's better to type the title as a caption in a report or as a slide title in a presentation.

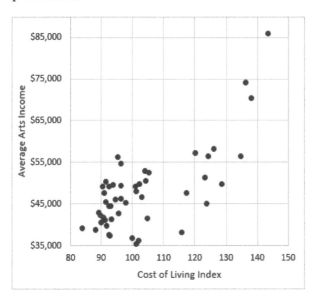

2. Save your Microsoft Excel workbook.

Screen print each chart

You can directly copy and paste Microsoft Excel tables and charts into Microsoft Word or PowerPoint files. For some destinations, however, such as the StoryMaps app, you'll need to capture images of charts and tables as files and insert them. You can use the Windows Snip & Sketch Tool, which is available on every Windows computer.

1. On your computer, search for the Snip & Sketch Tool, and start the program.

2. **Click New > Snip Now.**

3. **Drag a rectangle around your chart in Excel.**

4. **In the Snipping tool, click the Save As button, browse to your Chapter3\Tutorials folder, and for File name, type ScatterPlot. Make sure the Save As type is.jpg, and click Save.**

5. **Close Excel and the Snip & Sketch tool.**

Assignments

This chapter has four assignments to complete that you can download from ArcGIS Online, at go.esri.com/GISTforPro2.8Data:

- **Assignment 3-1**: Build a layout with income versus educational attainment in Washington, DC.
- **Assignment 3-2**: Build a map layout for comparing 2D and 3D maps for urgent health care clinics in Pittsburgh.
- **Assignment 3-3**: Build a StoryMaps story for locating charter schools in New York City.
- **Assignment 3-4**: Build a StoryMaps story for income versus educational attainment in Washington, DC.

PART 2

Working with spatial data

4

File geodatabases

LEARNING GOALS

- Import data into file geodatabases.
- Modify attribute tables and fields.
- Use Python expressions to calculate fields.
- Join tables.
- Get an introduction to SQL query criteria.
- Carry out attribute queries.
- Aggregate point data to polygon summary data.

Introduction

A database is a container for all the data of an organization, project, or other undertaking for record keeping, decision-making, analysis, or research. A file geodatabase is Esri's simplified database for storing geospatial data, including feature classes and raster datasets, and for processing by single users or small groups. In terms of data format, a file geodatabase is a file folder that has *.gdb* at the end of its name and contains files. A *.gdb* is not a file extension but just part of the folder name. In ArcGIS Pro, data processing in a file geodatabase is done through the Catalog pane, geoprocessing tools, and user interface.

For example, FoodDesertsChicago.gdb could be the name of a file geodatabase that stores spatial data for analyzing residents' access to grocery stores in Chicago, starting with feature classes for grocery stores, streets, and population in census blocks in Chicago. The workflow for estimating the number of Chicago residents who live in food deserts (that is, live a mile or more from the nearest grocery store) would have you create additional feature classes and tables, stored in FoodDesertsChicago.gdb. Chapter 9 provides the workflow and spatial analysis tools (buffers or service areas of grocery stores) for carrying out such a workflow.

Although file geodatabases have a simple format, they are powerful spatial data containers. For example, file geodatabases have no practical limits on numbers and sizes of feature classes and raster datasets stored in them; they are optimized for data processing and storage in ArcGIS Pro, and they allow data tables to be related and joined (essential database processes).

In this chapter, you will learn to work with file geodatabases in the following ways:

- Import spatial data into file geodatabases.
- Modify data.
- Join tables.
- Aggregate data on individual points to group-level data in polygons.
- Query data using attribute criteria.
- Build a spatial join (which is unique to GIS).
- Use the Catalog pane in ArcGIS Pro for utility work on file geodatabases.

In this chapter, you must work tutorials 4-1 through 4-3 sequentially, because tutorial 4-2 uses results from tutorial 4-1, and tutorial 4-3 uses results from tutorial 4-2.

Tutorial 4-1: Import data into a file geodatabase

When you create an ArcGIS Pro project, the software automatically creates a file geodatabase for you as the project's default spatial data container. In this tutorial, you'll create a project and import data from external sources into the new project's file geodatabase.

Create an ArcGIS Pro Project

1. **Open ArcGIS Pro.**

2. **Under New Blank Templates, click Map, and type or make the following selections. Clear the Create A New Folder For This Project check box and click OK.** When creating a project, a folder is often useful to hold the project file, geodatabase, toolbox, and so on along any additional files you may want associated with the project. As part of developing the exercise data for you, the Chapter4 folder was already created, so you won't need one in this case.

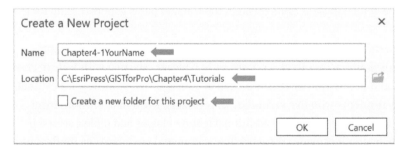

ArcGIS Pro creates the new project, Chapter4-1YourName.aprx, as well as a Chapter4-1YourName.gdb file geodatabase and Chapter4-1YourName.tbx toolbox. Every new feature class, table, and file that you create is saved by default to the project's file geodatabase—in this case, Chapter4-1YourName.gdb.

3. **In ArcGIS Pro, on the Project tab, click Options.** In the Options window on the Current Settings tab, you can see that the project, Tutorial4-1YourName, has Chapter4\Tutorials as its home folder, and its default geodatabase is the new Chapter4-1.gdb.

> **Change settings for the current project**
>
> Name
>
> Chapter4-1YourName
>
> Location
>
> C:\EsriPress\GISTforPro\Chapter4\Tutorials\Chapter4-1YourName.aprx
>
> Home folder
>
> C:\EsriPress\GISTforPro\Chapter4\Tutorials
>
> Default geodatabase
>
> C:\EsriPress\GISTforPro\Chapter4\Tutorials\Chapter4-1YourName.gdb
>
> Default toolbox
>
> C:\EsriPress\GISTforPro\Chapter4\Tutorials\Chapter4-1YourName.tbx

4. **Click Cancel, and click the back button.**

Set up a folder connection

When you created the Chapter4-1.aprx project file, ArcGIS Pro also created a folder connection to the folder in which you created the project (Chapter4\Tutorials). It's also useful to also have a connection to the Data folder (Chapter4\Data), which has shapefiles and other spatial data that you'll want to access conveniently (without having to browse through folders to find a location that you will often need). You will create a connection in the Catalog pane.

1. **Open the Catalog pane.**

2. **Expand Folders, and click Tutorials.** Here, you will see the contents of the Tutorials folder except for the project files (such as Tutorial4-1YourName.aprx). When you need to add data in the future, you can click the Tutorials folder and have direct access to your project's file geodatabase.

3. **Right-click Folders, and click Add Folder Connection.**

4. **Browse to EsriPress\GISTforPro\Chapter4, click the Data folder, and click OK.** ArcGIS Pro adds the Data folder connection.

5. **In the Catalog pane, expand the Data folder.** Here, you will see three folders and a geodatabase.

Import shapefiles and a data table into a file geodatabase

A shapefile is an Esri spatial data format, dating to the 1990s, for a single map layer. Although no longer Esri's preferred spatial data format, shapefiles are still used by many spatial data suppliers. For example, the US Census Bureau is the source of shapefiles used in this exercise. You will learn more about shapefiles and other spatial data formats in chapter 5, but for now, understand that ArcGIS Pro allows you to convert shapefiles and other map layer formats into feature classes in file geodatabases. When stored in a file geodatabase, it is called a feature class. When displayed as part of a map, it's a feature layer.

1. **On the Analysis tab, click Tools, and in the Geoprocessing pane, search for and open the** Feature Class To Feature Class **tool.** This tool converts a shapefile, coverage feature class (one of Esri's earliest spatial data formats), or geodatabase feature class to a shapefile or geodatabase feature class.

2. **For Input Features, click the Browse button.**

3. **In the Input Features window under Project, click Folders, double-click the Data folder, double-click the MaricopaCounty folder, select tl_2010_04013_cousub10.shp, and click OK.**

4. **For Output Name, type** Cities and run the tool. The input shapefile, for county subdivisions (cities, towns, and so on), was downloaded from census.gov, as was tl_2010_04013_cousub10.shp (census tracts). The new feature class, Cities, is automatically stored to the default file geodatabase, Chapter4-1YourName.gdb.

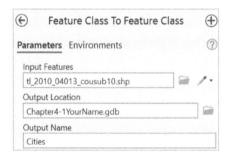

The imported feature class is automatically added to your map as the Cities map layer.

5. **Open the Catalog pane, and expand Databases > Chapter4-1YourName.gdb to see that the feature class, Cities, was created.**

6. **On the Catalog pane, click the Auto Hide button, and click the Auto Hide button again to hide.**

> **YOUR TURN**
>
> Import shapefile tl_2010_04013_tract10.shp from Chapter4\Data\MaricopaCounty to create the
> feature class, Tracts, in Chapter4-1YourName.gdb. Save your project.

Import a data table into a file geodatabase

Next, you'll import a data table with some 2010 census tract data for Maricopa County. The table
is in a common format for data sharing, comma-separated values, or.csv file. The first row of
the table has attribute names, and subsequent rows have data with values separated by commas
to delineate columns. GEOid is the Census Bureau's ID number for census tracts, and the other
attributes are census data. A nonprinting line-feed character is at the end of each row, signaling
the end of the row.

The figure shows some rows of the table you are about to import. Note that if the table you want
to import is in a Microsoft Excel workbook, you can save it as a.csv file and import that. Comma-
separated values data has the advantage of being easily imported into ArcGIS Pro; whereas,
depending on how your computer is set up, Excel workbook tables may not be recognized by
ArcGIS Pro.

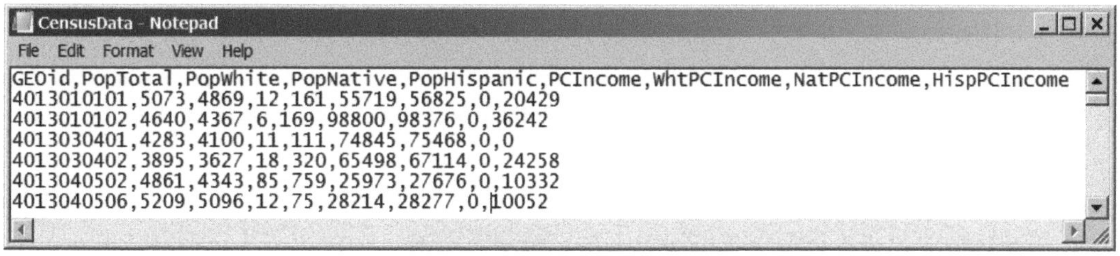

```
CensusData - Notepad                                               _ □ ×
File  Edit  Format  View  Help
GEOid,PopTotal,PopWhite,PopNative,PopHispanic,PCIncome,WhtPCIncome,NatPCIncome,HispPCIncome
4013010101,5073,4869,12,161,55719,56825,0,20429
4013010102,4640,4367,6,169,98800,98376,0,36242
4013030401,4283,4100,11,111,74845,75468,0,0
4013030402,3895,3627,18,320,65498,67114,0,24258
4013040502,4861,4343,85,759,25973,27676,0,10332
4013040506,5209,5096,12,75,28214,28277,0,10052
```

1. **Search for and open the** Table To Table **tool.**

2.   **For Input Rows, browse to Chapter4\Data\MaricopaCounty, select CensusData.csv, and click OK. For Output Name, type** CensusData and run the tool.

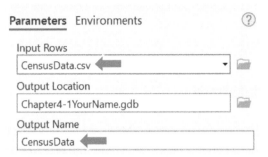

Use database utilities of the Catalog pane

You can create, copy, rename, and delete file geodatabases in the Catalog pane. You can do the same with feature classes and tables. To share the Tracts feature class and CensusData table with others, you'll create a file geodatabase for sharing, copy the spatial data you want to share, and rename the feature class and table.

1.   **Open the Catalog pane and expand Databases to see Chapter4-1YourName.gdb.**

2.   **Right-click Databases, and click New File Geodatabase. Browse to the Chapter4\Tutorials folder.**

3.   **For Name, type** MaricopaTracts, **and click Save.** MaricopaTracts.gdb is added to Databases.

4.   **Expand Chapter4-1YourName.gdb, right-click its Tracts feature class, and click Copy.**

5.   **Right-click MaricopaTracts.gdb, click Paste, and expand MaricopaTracts.gdb.** You can see that MaricopaTracts now has a copy of the Tracts feature class.

6.   **Copy the CensusData table from Chapter4-1YourName.gdb to MaricopaTracts.gdb.**

7.   **Right-click the Chapter4-1YourName.gdb and click refresh.** You will often need to refresh your geodatabase to see new files that you have added.

8.   **In MaricopaTracts.gdb, right-click CensusData, and rename it** TractsIncomeData. Now, you have just the tracts feature class and corresponding data table repackaged in the MaricopaTracts.gdb file geodatabase, which you can share with others. You don't need this file geodatabase for any further processing in this chapter, so next you'll delete it.

9. In the Catalog pane, right-click **MaricopaTracts.gdb**, click **Delete**, and click **Yes**. Deleting tables and features from a file geodatabase is permanent versus removing a layer from Contents, where the feature class still resides in a file geodatabase and can be added again later.

10. **Save your project.**

Tutorial 4-2: Modifying an attribute table

Much of what gets processed or displayed in a GIS depends on attributes—columns in tables. To get tables in the desired form, you must know how to delete, create, and modify attributes.

Get started

1. **Save your Tutorial4-1YourName project as** Tutorial4-2YourName **in Chapter4\Tutorials.**

2. **In the Contents pane, right-click Tracts, and click Zoom To Layer.**

3. **Create a bookmark named** Maricopa County.

4. **Symbolize Tracts and Cities with no color for the fill and Cities with a wider and darker gray outline than Tracts.**

5. **Change the basemap to Light Gray Canvas, and remove the layer World Light Gray Reference from the Contents pane.**

Delete unneeded columns

Many feature classes have extra or unnecessary attributes from the user's point of view that you do not need and can delete.

1. **In the Contents pane, right-click Tracts, and click Design > Fields.** This view allows you to modify and create attributes in a table. You cannot delete the primary key, ObjectID, or the Shape, Shape_Length, and Shape_Area attributes because they are essential. These fields are dimmed, meaning that you can't modify them. GeoID10 is the tract geocode that you need later in this chapter for joining the CensusData table to the tracts map layer, but all other fields are candidates for deletion for the purposes of this chapter.

2. **Press the Ctrl key while selecting rows crossed out in the figure, right-click in the group, and click Delete. To delete multiple rows at once, hold down the Shift key, and click the first and last rows in the group. Hold the Ctrl key to deselect a field, and then right-click and click delete.** ArcGIS Pro crosses out the fields but does not delete them until you click Save. Review selections for deletion before saving. If you decide to keep any of these fields, you can right-click, and click Restore Deleted Item.

◢	☑ Visible	■ Read Only	Field Name	Alias	Data Type
	☑	☑	OBJECTID	OBJECTID	Object ID
	☑	☐	Shape	Shape	Geometry
	☑	☐	~~STATEFP10~~	~~STATEFP10~~	~~Text~~
	☑	☐	~~COUNTYFP10~~	~~COUNTYFP10~~	~~Text~~
	☑	☐	~~TRACTCE10~~	~~TRACTCE10~~	~~Text~~
	☑	☐	GEOID10	GEOID10	Text
	☑	☐	~~NAME10~~	~~NAME10~~	~~Text~~
	☑	☐	~~NAMELSAD10~~	~~NAMELSAD10~~	~~Text~~
	☑	☐	~~MTFCC10~~	~~MTFCC10~~	~~Text~~
	☑	☐	~~FUNCSTAT10~~	~~FUNCSTAT10~~	~~Text~~
	☑	☐	~~ALAND10~~	~~ALAND10~~	~~Double~~
	☑	☐	~~AWATER10~~	~~AWATER10~~	~~Double~~
	☑	☐	~~INTPTLAT10~~	~~INTPTLAT10~~	~~Text~~
	☑	☐	~~INTPTLON10~~	~~INTPTLON10~~	~~Text~~
	☑	☑	Shape_Length	Shape_Length	Double
	☑	☑	Shape_Area	Shape_Area	Double

3. **On the Fields tab, click Save.**

4. **Close Fields view. Open the Tracts attribute table.** Only the needed fields remain.

◢ OBJECTID	Shape	GEOID10	Shape_Length	Shape_Area
1	Polygon	04013422644	0.127403	0.000755
2	Polygon	04013422643	0.09817	0.000499
3	Polygon	04013422642	0.063555	0.000249
4	Polygon	04013422641	0.132432	0.000748
5	Polygon	04013815900	0.170811	0.001141
6	Polygon	04013815800	0.075694	0.000274

YOUR TURN

Delete all fields from Cities except for the dimmed fields, GEOID10, and NAME10. While still in the Fields view, type **City** for NAME10's alias. Open the Cities attribute table and see the updated fields and alias for Name10.

Examine geocodes in tables to be joined

In the next exercise, your work will prepare for joining a data table (CensusData) to the attribute table of the Tracts feature class on a one-to-one basis: each row or record in the CensusData table has one matching record in Tracts. Joining two one-to-one tables makes a wider table (increases the number of columns) while leaving the number of rows the same. This is common in GIS. Joining two tables together requires each table to have an attribute with matching values stored as the same data type (for example, numeric or text). For this census tract data, the matching values are census tract geocodes, GEOID10 in Tracts and GEOid in CensusData.

To make a map using census data, you must download a table with the census tract attributes you are interested in from the Census Bureau website (see chapter 5) and then download the geometry of the census tracts. You must then join the table with the attributes of interest to the tracts attribute table based on the census tract ID field that each file has. Joining tables involving census data is more practical than storing the thousands of census tract attributes in the Tracts attribute table. The attribute table for Tracts, as downloaded from the Census Bureau's website, contains the tract geocode and a few GIS attributes including the tract geometry but no census attributes.

The geocodes in CensusData and Tracts could be joined, except GEOID10 in Tracts has a text data type whereas GEOid in CensusData has a numeric data type. The remedy is to create a new field in Tracts, name it something like GEOID10Num (to indicate the data type is numeric), set the data type to be numeric, and copy the data from the GEOID10 field.

You can drop leading zeros from tract geocodes that are stored as text data type when converting them to a numeric data type. Because GEOid is numeric, it has leading zeros already dropped, so one recourse for getting matching values is to drop the leading zeros in GEOID10 by transferring its data into a numeric data field.

1. **In the Contents pane, right-click CensusData, click Design, and click Fields.** You can see that GEOid has a numeric data type, Double.

2. **Close the CensusData Fields view, and open the Tracts Fields view.** GEOID10 has the text data type.

3. **Close the Tracts Fields view, open the Tracts attribute table, and sort GEOID10 ascending.** The first row's value for GEOID10 has the text value 04013010101.

4. **Open the CensusData table and sort GEOid ascending.** CensusData has a corresponding numeric value, 4013010101, without the leading zero. If you created a text attribute in CensusData named GEOidText, its corresponding value would remain 4013010101, without the leading zero. You could add the leading zero, but it is easier to create a numeric attribute in Tracts that automatically drops any leading zeros so the geocodes match.

Create an attribute and populate it using the Calculate Field tool

Next, you'll create a numeric attribute in the Tracts table and use the Calculate Field tool to transform and copy data to it.

1. **With the Tracts attribute table open, click its Options button (top right) and select Fields view.**

2. **At the bottom of Fields view, click Here To Add A New Field.**

3. **Type** GEOID10Num **for the Field name, and select Double as its data type.**

4. **On the Fields tab, click Save.**

5. **Close Fields view.** Notice that the values for GEOID10Num are <Null>, which signifies that the data cells are empty and no data has been added yet. These values are called null values.

6. **Right-click GEOID10Num in the Tracts table, and click Calculate Field.**

7. **In the Calculate Field pane, double-click GEOID10 in the Fields panel to create the expression GEOID10Num = !GEOID10!.** The expression's syntax is from the Python programming language, requiring attribute names on the right side to be identified using the exclamation point delimiters.

8. Run the tool.

9. **View the results in the attribute table.** The new attribute, GEOID10Num, has GEOID10 values transformed to numeric format without the leading zeros and will match perfectly with the geocode in CensusData.

OBJECTID *	Shape *	GEOID10	Shape_Length	Shape_Area	GEOID10Num
34	Polygon	04013010101	0.493324	0.005285	4013010101
644	Polygon	04013010102	3.412224	0.266627	4013010102
662	Polygon	04013030401	0.261295	0.002584	4013030401
643	Polygon	04013030402	0.411925	0.00567	4013030402
86	Polygon	04013040502	0.229298	0.002005	4013040502

10. **Close the Tracts table.**

YOUR TURN

Next, you'll use data in the CensusData table to make a choropleth map for Maricopa County that compares incomes of Hispanic and Native American people to those of White people. The CensusData table has population and per capita income for the total population, White, Native American, and Hispanic. You will map two ratios of per capita income for Hispanic and Native American people divided by per capita income for White people. To get started, create two attributes in the CensusData table, **RatioHispanicWhiteIncome** and **RatioNativeWhiteIncome**, both with the float data type. When finished, save and close the Fields view.

Calculate the ratio of two fields

When you calculate ratios, ensure that you do not divide by zero, which is an undefined operation in mathematics. In ArcGIS Pro, you can select rows in which the divisor is not zero. The Calculate Field tool will include only the selected rows, thus avoiding division by zero.

1.   **Open the CensusData table.**

2.   **On the Map tab in the Selection group, click the Select By Attributes button** .

     Next, you'll create the condition needed for selecting table rows in which the divisor, WhtPCIncome, is greater than 0. In tutorial 4-4, you'll learn a lot more about attribute queries.

3.   **In the Select By Attributes pane, click New Expression, make the following selections as shown and run the tool.** You can click the Values drop-down list, expanding the list to display all existing values in the WhtPCIncome field, and then click a value (such as the zero needed here) to complete the query criterion expression. Using the drop-down list will correctly format data values for use in queries, which works well for date and text attributes.

| Where | WhtPCIncome ▾ | is greater than ▾ | 0 ▾ | ✕ |

     ArcGIS Pro selects 908 of 913 tracts, all with positive values for WhtPCIncome.

4.   **Right-click the heading for the RatioNativeWhiteIncome field, and click Calculate Field.**

5.   **Create the expression as seen in the figure by double-clicking fields and clicking the division operator. Run the tool.**

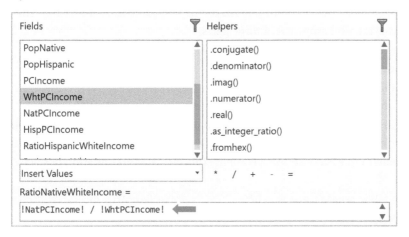

     If you get an error message, it's likely that you don't have records selected with nonzero WhtPCIncome.

6. **In the table, right-click RatioNativeWhiteIncome, and click Sort Descending.** The figure has the results for six rows sorted by RatioNativeWhiteIncome. In some tracts, Native American income is more than four times higher than White income, but in most tracts, the ratios are less than 1.

RatioNativeWhiteIncome ▾
4.605406
3.80295
3.5819
3.462956
3.394832
2.919134

YOUR TURN

With the 908 rows of CensusData with positive WhtPCIncome still selected, create and run the following expression:

```
RatioHispanicWhiteIncome = !HispPCIncome! /
!WhtPCIncome!.
```

In the figure, you will see the first six records, sorted descending for the results.

When finished, clear the selection, close the table, and save your project

RatioHispanicWhiteIncome ▾
2.073832
2.016204
1.904018
1.79416
1.691166
1.547541

Extract substring fields and concatenate string fields

If you sort the Tracts attribute table ascending for GEOID10, the first value is 04013010101. This unique identifier for a census tract is composed of three parts:

- 04 is the ANSI (American National Standards Institute) code for Arizona.
- 013 is the ANSI code for Maricopa County.
- 010101 is the ANSI identifier for a census tract.

The census tract is further broken down to tract number: the first four digits (0101) and an optional two-digit suffix (01) with an implied decimal point between the number and suffix. Formally written, the corresponding census tract name for Maricopa County, Arizona, is Census Tract 0101.01. Using GEOID10 as the input, in this exercise you'll compute this formal representation of tracts.

1. **In the Tracts attribute table, create three text fields:** TractNumber **(length = 4),** TractSuffix **(length = 2), and** TractName **(length = 20).**

2. **Calculate the TractNumber field with the expression** `!GEOID10![5:9]`. To understand the Python language syntax of this expression, consider the case of tract 04013123456. The characters stored (digits, in this case) of any field are indexed 0, 1, 2, ..., from left to right, so the positions of the 04 for Arizona are 0 and 1. The notation [0:2] extracts the indexed first digit and up

to but not including the third digit, 0. So 04 is extracted. For the case of extracting TractNumber and the example of 04013123456, the 5 of !GEOID10![5:9] corresponds to the sixth position (1) and the 9 to the 10th position (5), which is not included. So [5:9] extracts 1234.

3. **Calculate the TractSuffix field with the expression** `!GEOID10![9:11]`. This expression extracts the ninth digit up to but not including the 11th character and produces the value 01.

4. **Finally, calculate TractName with the following expression:**

```
"Census Tract " + !TractNumber! + "." + !TractSuffix!
```

Note that there is a space after Census Tract and before the closing double quotation mark. The plus sign (+) concatenates or combines two text values into one. "Census Tract " and "." in double quotation marks are constants, the same for every row of data. !TractNumber! and !TractSuffix! with exclamation point delimiters denote field names, which vary by row. Sorted ascending by TractName, the figure shows the first six finished values.

TractNumber	TractSuffix	TractName ▲
0101	01	Census Tract 0101.01
0101	02	Census Tract 0101.02
0304	01	Census Tract 0304.01
0304	02	Census Tract 0304.02
0405	02	Census Tract 0405.02
0405	06	Census Tract 0405.06

YOUR TURN

An example value for GEOID10 in the Cities attribute table is 0401390459. The first two digits are the state, the next three digits are the county, and the remaining five digits are the city. In the Cities attribute table, create two text attributes, CityNumber (length = 5) and CityNameNumber (length = 50). Extract the last five digits of GEOID10 to populate the CityNumber field,

City	CityNumber	CityNameNumber ▲
Buckeye	90459	Buckeye = 90459
Chandler	90561	Chandler = 90561
Deer Valley	90867	Deer Valley = 90867
Gila Bend	91377	Gila Bend = 91377
Phoenix	92601	Phoenix = 92601
Salt River	93060	Salt River = 93060

and then concatenate the City (Name 10) field and CityNumber to compute values for CityNameNumber, such as Buckeye = 90459. Put a space on both sides of the equals sign. The concatenation syntax is as follows: **!NAME10! + " = " + !CityNumber!**. Sorted ascending by CityNameNumber, sample values are shown. Save your project.

Tutorial 4-3: Joining tables

The table join you will do next adds the columns of the CensusData table to the attribute table of Tracts by matching values of GEOid from the CensusData table to GEOID10Num from Tracts. In this one-to-one join, each record of both tables finds a single matching record.

Each time you open a project with a table join, the join is re-created on the fly. To make a permanent Tracts and Census Data table join, you can export the Tracts feature class to a new feature class. Then the joined attributes are permanent in the new feature class. If you want to publish a feature class with a table join in ArcGIS Online, you must make the join permanent.

Get started

1. **Save your Tutorial4-2YourName project as** Tutorial4-3YourName **in Chapter4\Tutorials.**

2. **Use the Maricopa County bookmark.**

Join a data table to a feature class attribute table

Next, you will join the CensusData table to the Tracts feature class.

1. **In the Contents pane, right-click Tracts, and click Joins and Relates > Add Join.**

2. **In the Geoprocessing pane, for Input Join Field, select GEOID10NUM. Run the tool.**

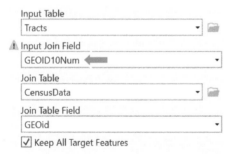

3. **Open the Tracts attribute table, scroll to the right in the table, and verify that ArcGIS Pro joined the CensusData table to the Tracts attribute table.**

4. **Close the Tracts table.**

Export a feature class to make a join permanent

This exercise turns the join attributes of CensusData into permanent attributes of the Tracts attribute table by exporting the Tracts feature class. Also, you'll clean up the attributes of the exported feature class.

1. In the Contents pane, right-click Tracts, and click Data > Export Features.

2. In the Geoprocessing pane, change the Output feature class's name to MaricopaIncome, **and run the tool.**

3. Remove Tracts from the Contents pane.

YOUR TURN

Delete the following fields from the MaricopaIncome attribute table: GEOID10, OBJECTID, and GEOid. Now you can answer the question: Are there places in Maricopa County where Hispanic people typically have higher per capita income than White people? Symbolize MaricopaIncome with graduated colors, set RatioHispanicWhiteIncome as the field, Standard Deviation for method, 1 standard deviation for interval size, and red to yellow to green for color scheme and a medium gray outline. Label Cities, and use a Light Gray Canvas base. You can see many places throughout Phoenix and Chandler where Hispanic people earn more than White people. The tracts that have a white color are the ones for which no data was available on income for Hispanic people. Save your project.

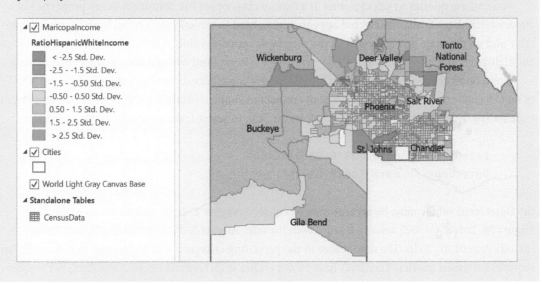

Tutorial 4-4: Attribute queries

One of the major innovations of GIS technology nearly 50 years ago was linking tabular data to the spatial features in feature classes. This linkage allows you to symbolize maps using the attribute values found in tables and ensures that attribute queries show only features of interest.

Attribute queries are based on Structured Query Language (SQL), the de facto standard query language of database packages and many application software packages, including ArcGIS Pro. This tutorial introduces the query criteria part of SQL. For further study and to learn full SQL commands, you can find free interactive SQL tutorials on the internet. A good one is at w3schools.com. The next two tutorials use crime data and queries commonly made by police crime analysts. The example queries ask some of the standard query questions of "who," "what," "where," "when," and "how."

A simple SQL criterion has the following form:

```
attribute name <logical operator> attribute value
```

The attribute name can be any attribute column heading or field name in an attribute table. Several logical operators are available, including the familiar ones such as =, >, >=, <, and <=. The attribute value is related to the values you seek. For example, the following simple criterion selects all crimes that are robberies where Robbery is a value of the Crime attribute:

```
Crime = 'Robbery'
```

Text values, such as Robbery, must be enclosed in single or double quotation marks.

You can use queries to select points in a feature class or set the definition query properties of a feature class. A point that is selected appears highlighted in the selection color, both in the attribute table and on the map, and all points of the map layer remain visible. A definition query, on the other hand, displays only the points that satisfy the query criteria (and without the need for highlighting).

Compound criteria are made up of two or more simple criteria connected with either an AND or an OR connector. AND means that both connected simple criteria must be true for corresponding features (records) to be selected. For example, for SQL to select features for the compound query,

```
DateOccur >= date '2016-08-01' AND
DateOccur <= date '2016-08-28',
```

the DateOccur values must be greater than or equal to August 1, 2016, and less than or equal to August 28, 2016. SQL will select a feature with the date August 5, 2016, for example, but it will exclude August 31, 2016. The date values in the preceding criteria are in the format that ArcGIS Pro requires for use in queries. You don't have to remember such formats because the ArcGIS Pro query builder helps you get the values (and in the right format) you need in queries, as you will see in the next tutorial.

The OR connector is used to select a subset of features on the basis of classification, such as crime types in the Crime attribute. Although a single criminal incident may have one or more criminal offenses (for example, homicide and robbery in which a person was robbed and murdered), data reported by police departments to the FBI has only the most serious offense reported (homicide, in this case), so each crime has only one reported crime type. To query this kind of data for homicides and aggravated assaults, you can simply say, "I want all homicides and aggravated assaults," but in SQL, the expression is "homicides OR aggravated assaults." Given FBI reporting practices, it is impossible for an offense to be both a homicide and an aggravated assault because only one crime type is recorded for each offense. So, the proper SQL criteria in this case are:

```
Crime = 'Homicide' OR
Crime = 'Aggravated Assault'.
```

Despite occasional exceptions to this rule, you will almost always need to place OR in parentheses (as shown) if the criteria are combined with other criteria, such as a date range:

```
DateOccur >= date '2015-08-01' AND
DateOccur <= date '2015-08-28' AND
(Crime = 'Homicide' OR
Crime = 'Aggravated Assault').
```

The use of parentheses (just as with algebraic expressions) is because logical expressions are executed one pair at a time for simple expressions, generally working from left to right but with certain logical operators going first. For example, SQL executes AND comparisons before OR comparisons regardless of order, which can result in incorrect information unless you use parentheses to control the order of execution. SQL executes comparisons in parentheses first. For example, if you leave the parentheses out of the previous query, you'd retrieve all homicides in August 2015 but also all aggravated assaults in the database regardless of the month. Verifying input data and the output results will give you experience and guidance in making compound queries.

Police need three primary kinds of attribute queries to analyze crimes, and you can combine and use these queries creatively in the following ways:

- The most fundamental attribute query is by crime type and data interval (the *what* and *when*). Such queries often combine several crime types with the use of logical operators.
- The second type of attribute query adds criteria, such as time of day or day of the week—for example, weekday versus weekend crime (a refinement of *when*). Clearly, crime patterns can differ as a result of the time of day or the day of the week.
- The third type of attribute query adds criteria based on the attributes of the people (the *who*) or the objects (the *what*), such as vehicles, involved in a crime and can include modus operandi attributes (the *how*), such as "Enters through an open window," if available.

Open the Tutorial 4-4 project

1. **Open Tutorial4-4YourName in Chapter4\Tutorials, and save it as** Tutorial4-4YourName. **Use the Pittsburgh bookmark.** The map of Pittsburgh has Crime Offenses, which has all criminal offenses for June through August 2015, Streets, and Neighborhoods. Note that the crime data, although real, has been modified to protect privacy, including changing the year of crimes. Also note that Pittsburgh is a relatively low-crime city, and that crime data plotted in any city would likely look like the crime data you will plot in this tutorial.

Create a date-range selection query

Queries for event locations, such as crimes, almost always use date-range criteria. A selection by attributes identifies the subset of features in a map layer that meet the query criteria and gets the selection color both in the table and on the map. Once you select the features, you can do additional processing of selected outputs. For example, you can export the selected features to a new feature class.

1. **On the Map tab, in the Selection group, click Select By Attributes.**

2. **In the Geoprocessing pane, for Input Rows, select Crime Offenses, click New Expression, and make the query Where DateOccur is on or after 7/1/2015 12:00:00 AM, as shown. Ensure that you click in the cell with the date, and select the date from the drop-down list.**

3. **Click the Verify button** ✓. The Verify button ensures that there are no syntax errors but does not check for logic errors.

4. **Click Add Clause, and make the query Where DateOccur is on or after 7/1/2015 12:00:00 AM And DateOccur is on or before 7/31/15 12:00:00 AM, as shown.**

| Where | DateOccur | ▾ | is on or after | ▾ | 7/1/2015 12:00:00 AM | ▾ | 🗓 | ✕ |
| And ▾ | DateOccur | ▾ | is on or before | ▾ | 7/31/2015 12:00:00 AM | ▾ | 🗓 | ✕ |

5. **Click the Verify button.**

6. **Click Apply, and leave the Geoprocessing pane open.** All crime offenses remain on the map, but the points for July 2015 are now displayed in the selection color. Next, you'll save the query so that it can be reused.

7. **In the Geoprocessing pane, click the Save Expression button** 💾**, and save the expression as** qryDateRange **in Chapter4\Tutorials.** In the next exercise, you'll reload this query for a definition query. The "qry" prefix is a standard prefix for database queries.

8. **Open the Crime Offenses attribute table, and click the Show Selected Records button.** Notice that 3,924 out of 11,500 features are selected.

9. **Close the attribute table and the Select by Attributes pane.**

10. **Clear the selection.**

Reuse a saved query to create a definition query

1. **Right-click Crime Offenses, click Properties > Definition Query, click the New Definition Query drop-down menu, select Add Definition Queries From Files, and double-click qryDateRange.exp.** The saved expression reloads.

2. **Click OK.**

3. **Open the Crime Offenses attribute table, click Load All, and verify that only the 3,924 July 2015 crimes remain.**

4. **Close the table.**

Query a subset of crime types using OR connectors and parentheses

Next, you'll add clauses to query a subset of crimes that are leading indicators of burglaries. A spatial cluster of disorderly conduct, vagrancy, or vandalism crimes in July usually will indicate an increase in the more serious crime of burglary nearby in August. In cases in which you need a subset of classes (in this case, crime types), you will select crimes using the OR connector with the connected OR clauses in parentheses. OR conditions can be difficult so always view the output of queries you build for the first time to ensure that you are getting the desired information.

1. **Right-click Crime Offenses, and click Properties > Definition Query > Edit to open the Definition Query property sheet for Crime Offenses.** Your two date clauses connected with AND are still there. You'll add three clauses, one for each leading-indicator crime, connected by ORs, and type parentheses around them using the SQL view of the query.

2. **Click Add Clause, and add the clause And Crime is equal to Disorderly Conduct, as shown.**

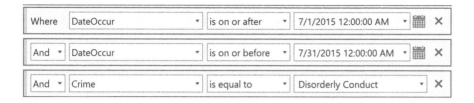

3. **Click Add Clause.**

4. **Add two more clauses for Crime Equal To Vagrancy and Crime Equal To Vandalism, connected by OR.** If you ran this query without parentheses around the OR clauses, you'd get disorderly conduct features in July 2015 and then all vagrancy and vandalism in the dataset (not just in July).

5. **Click the SQL button.** This step shows you the actual SQL criteria that the query builder built. The SQL is mostly self-explanatory except for the values in red, which are now formatted as needed for SQL. You rarely need to type values, but instead select them from drop-down lists.

```
DateOccur >= timestamp '2015-07-01 00:00:00' And DateOccur <= timestamp ↵
'2015-07-31 00:00:00' And Crime = 'Disorderly Conduct' Or Crime =      ↵
'Vagrancy' Or Crime = 'Vandalism'
```

6. **Type an opening parentheses just before the first "Crime" and a closing parenthesis at the very end, as shown.** The SQL first will evaluate the expression inside the parentheses, which is to select all disorderly conduct, vagrancy, or vandalism crimes. Next, SQL will work from left to right to select all records in July for the three crime types.

```
DateOccur >= timestamp '2015-07-01 00:00:00' And DateOccur <= timestamp ↵
'2015-07-31 00:00:00' And (Crime = 'Disorderly Conduct' Or Crime =      ↵
'Vagrancy' Or Crime = 'Vandalism')
```

7. **Click Apply and click OK.** Open the Crime Offenses attribute table, and verify that the 724 remaining features are for the three crimes of interest in July.

8. **Open the definition query for Crime Offenses, hover over the query expression, click Save, and save the query as** qryDateRangeBurglaryLeadingIndicators. This definition query is useful, because if you need to query more deeply about the leading indicator crimes in July 2015, you can create selection queries starting with the map layer created by the definition query. Your selection queries will be simpler with the basic query already done by the definition query.

9. **Hover over the query expression, click Remove definition query, click Yes, and click OK.** Now you're back to having all original features displayed.

YOUR TURN

Create a saved definition query named **qryAugustBurglaries.exp** for burglaries in August 2015. Open the attribute table to verify the results. There are 273 records. Keep the definition query in effect (don't clear it). Activate each new query by clicking its green circle check mark.

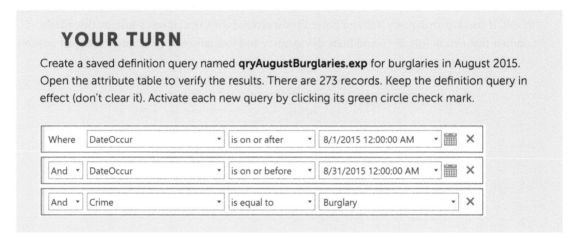

Query day-of-week range

Next, you'll use the attribute DayOfWeek in Crime Offenses, which identifies the day of the week for a crime (Monday through Sunday). You'll build a selection query for burglaries during the weekend (Saturday and Sunday) and get the number of weekend crimes in August. Then you'll switch the selection so that weekdays are selected and get the number of weekday crimes in August. You can also get the weekday total by subtracting the number of weekend burglaries in August from the total number of August burglaries, but you will learn about switching selections in ArcGIS for practice. Finally, you'll get the average number of burglaries per day on a weekday versus a weekend day. Which do you think should be higher?

1. **Create the Select By Attributes query (and not a definition query) Where DayOfWeek is equal to Saturday Or DayOfWeek is equal to Sunday, as shown.**

2. **Apply the query.**

3. **Save the query as** qryWeekend.

4. **Close the Geoprocessing pane.**

5. **Open the Crime Offenses attribute table.** Eighty-four of 273 burglaries are selected.

6. **In the attribute table, click the Switch Selection button** . You will see that 189 of the 273 features are weekday burglaries.

    August 2015 had 10 weekend days and 21 weekdays. Doing the math, an average of 8.4 burglaries occurred per weekend day and 9.0 occurred each weekday. Burglaries often occur when residents are away from home, so normally one expects weekdays to have more burglaries than weekends, although the difference is small in this case.

7. **Clear the selection, and close the table.** The definition query is still set to August Burglaries, so you have 273 of them mapped with none selected.

Query time of day

In this case, police work three shifts (with military times in parentheses):

1. Day: 7:00 a.m.–3:00 p.m. (0700–1500)
2. PM: 3:00 p.m.–11:00 p.m. (1500–2300)
3. Night: 11:00 p.m.–7:00 a.m. (2300–2359 and 0000–0700)

Now you can see how many August burglaries each shift had. You'll query for the PM and night shifts and get the day shift by calculating the difference. You will include the query for night shift as another example of a query requiring an OR connecter. The data has military time as the numeric attribute, TimeOccur, which ranges between 0 and 2359 (the kind of time data is a separate attribute that you need for working with time-of-day queries in ArcGIS). The two SQL clauses for night shift are TimeOccur >= 2300 OR TimeOccur < 700. These clauses work because the maximum of TimeOccur is 2359 and the minimum is 0. Normally, the right sides of intervals (such as the 700 in the second clause) are not included and use only less than. The 700 is included in the day shift.

1. **Start a Selection By Attributes query for Crime Offenses (now all August 2015 burglaries).** You can start with the night shift.

2. **Add the two night shift clauses** (TimeOccur >= 2300 OR TimeOccur < 700), **and apply the query.** Open the Crime Offenses attribute table, and see that 82 (30 percent) of 273 burglaries were during the night shift.

3. **Clear the query and table.**

4. **Build two clauses for the day shift, which in SQL are** TimeOccur >= 700 AND TimeOccur < 1500, **run the query, and open the table.** Ninety-eight (36 percent) of 273 are dayshift burglaries. That means that 273 − 82 − 98 = 93 (34 percent) are in the PM shift. So, although the maximum

(36 percent) are daytime burglaries while residents are at work or away, and the minimum (30 percent) are nighttime burglaries, following the expected pattern, the percentages do not vary much. Nevertheless, police find the results useful.

5.   **Clear the expression and selection and close the table.**

Query person attributes

Suppose an informant told a police officer that a burglary was committed by a white male in his 30s who lives on Warrington Avenue, and the officer wants to see if a person with those characteristics has already been arrested. The query will need several clauses all connected by AND because the person sought must have all the provided characteristics. Also, you must search for a part of text values, namely just the street name (Warrington) from an address (such as 123 Warrington Av). You can use SQL.

1.   **Open the Crime Offenses table, and start a Select By Attributes query.** This time you'll keep running the query as you build it and see how the query results are narrowed. Start by excluding all records in which there was no arrest, which you'll find by querying for the arrested person's last name as not being null. Null means no value entered, which, in this case, is the indicator for no arrest made.

2.   **Add a clause for ArrLName Is Not Null, run it, and click the Show Selected Records button on the lower-left corner of the table.** There were 40 arrests made for burglaries committed in August, all of them now visible and in the selection color. Of course, these include different genders, races, and ages.

3.   **Add and apply an AND clause for** `ArrSex = M`. Now there are 34 of 40 records selected, with evidently only six female burglars arrested.

4.   **Add and run an AND clause for** `ArrRace = W`. Slightly more than half the burglaries with arrests remain, 18 being White burglars.

5.   **Add and run two AND clauses for** `ArrAge >= 30` **and** `ArrAge < 40`. You're down to six White males in their 30s who were arrested for burglary. You can review the records themselves to visually apply the last criterion, street name, but for other cases this can be inefficient. So you'll add the last clause.

6.   **Add and run an AND clause for ArrResid "contains the text" Warrington, in which you type** Warrington. No one satisfies the last condition, but a bit of digging in the data reveals that street names are entered with all caps (for example, WARRINGTON), and SQL is case sensitive for values. You must edit the last clause.

7. **Point to the last clause, and type** WARRINGTON **for its value.** Editing a clause is useful when you must make several queries, such as modifying parts of the query as in assignment 4-3 (which you can download from ArcGIS Online).

Where	ArrLName ▾	is not null ▾		✕
And ▾	ArrSex ▾	is equal to ▾	M ▾	✕
And ▾	ArrRace ▾	is equal to ▾	W ▾	✕
And ▾	ArrAge ▾	is greater than or equal to ▾	30 ▾	✕
And ▾	ArrAge ▾	is less than ▾	40 ▾	✕
And ▾	ArrResid ▾	contains the text ▾	WARRINGTON ▾	✕

8. **Click the SQL button to see the command that you just built.** A new SQL logical operator is LIKE, which is used only on text-valued attributes. The value, WARRINGTON, is enclosed in single quotation marks so that SQL learns that it's a text value, and the % symbol is the wildcard character for ArcGIS SQL that stands for zero, one, or more characters to ignore. The record that you'll retrieve has the value 1005 E WARRINGTON AV for ArrResid. The first % in the query value ('%WARRINGTON%') ignores seven characters at the beginning of the address (1005 E), where the blank spaces are counted as characters, and the last % ignores three characters at the end (AV).

```
ArrLName IS NOT NULL And ArrSex = 'M' And ArrRace = 'W' And ArrAge >= 30
And ArrAge < 40 And ArrResid LIKE '%WARRINGTON%'
```

9. **Run the command.** You now have an arrested person who also may have committed the unsolved burglary, John Bond.

10. **Clear the selection, close the table, and save your project.**

Tutorial 4-5: Data aggregation with a spatial join

In this tutorial, you will count (aggregate) burglaries by neighborhood, providing the big picture of crime. Aggregation of point data requires a spatial join of burglary points to neighborhood polygons. The spatial join process can determine the polygon in which a point lies, enabling the data aggregation. You'll use the results to create a choropleth map of burglaries by neighborhood in August 2015. You can do all of this in ArcGIS Pro.

Open the Tutorial 4-5 project

1. **Open Tutorial4-5.aprx in Chapter4\Tutorials, and save it as** Tutorial4-5YourName.aprx.

2. **Use the Pittsburgh bookmark.** The map is the one that you used in tutorial 4-4 with the definition query yielding burglaries in August 2015. Neighborhood labels have been added and streets removed.

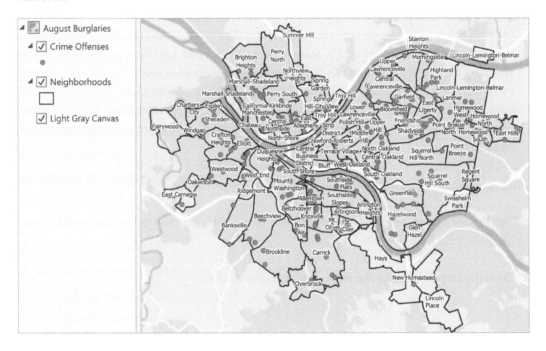

Build a spatial join

1. **Search for and open the** Spatial Join **tool.**

2. **Type or make the selections as shown. In the Output Fields list, starting with Area, point to each field and click the Remove button so that only Name (from Neighborhoods) remains and run the tool.** Notice that Neighborhoods, the polygon outputs you need, are Target Features. Make sure that Keep All Target Features is selected so that neighborhood polygons that don't have any burglaries within them will also be included in the output feature class. Polygons are output for neighborhoods that had no burglaries.

3. **Open the new map layer's attribute table.** See that Join_Count is added as a new attribute and is equal to the number of burglaries in each neighborhood.

4. **Close the table.**

YOUR TURN

Create a choropleth map using Join_Count from August2015BurglariesByNeighborhood. Use five quantiles and the color scheme of your choice. Remove labels for the August2015BurglariesbyNeighboood layer. Turn off the Crime Offenses layer. When finished, save your project.

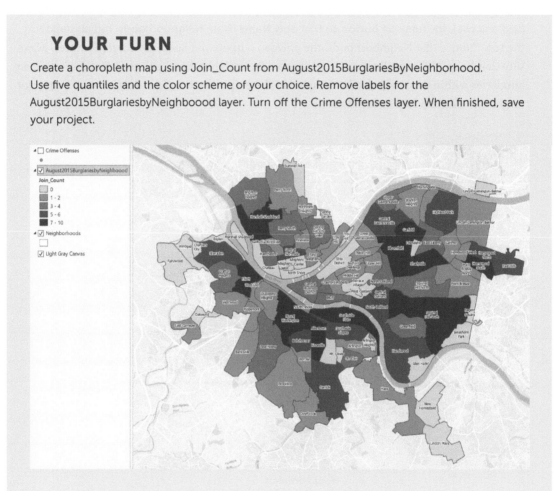

Tutorial 4-6: Central point features for polygons

If you choose graduated symbols for symbology of a polygon map layer, ArcGIS Pro creates the central points on the fly and renders them as point features. This choice works well and saves you the work of creating a new central point feature class from the polygons yourself. However, it is often helpful to create a separate feature class with central points created from polygon features, so you will create a separate feature class.

Get started

1.   **Open Tutorial4-6.aprx from Chapter4\Tutorials, and save it as** Tutorial4-6YourName.aprx**.**

2.   **Use the Pittsburgh bookmark.** This map has the aggregate number of burglaries by neighborhood in Pittsburgh that you created in tutorial 4-5, without the neighborhood labels.

Create a central point feature class for polygons

The centroid of a polygon is the arithmetic mean of all points within the polygon. For most polygons, centroids lie within their polygons, but for some, such as a quarter-moon-shaped polygon, centroids lie outside. If you want all center points to lie within their polygons, the remedy in ArcGIS is to use central points instead of centroids. Central points all lie within their polygons. In this tutorial, you'll use the Feature To Point tool to create a central point feature class. Although not necessary, for practice you'll first use the Add Geometry Attributes tool for adding central point coordinate attributes to the feature class. It's valuable to know about the Add Geometry Attributes tool.

1.   **Search for and open the** Add Geometry Attributes **tool.**

2.   **For Input Features, select BurglariesByNeighborhood. For Geometry Properties, select Central point coordinates. Run the tool.**

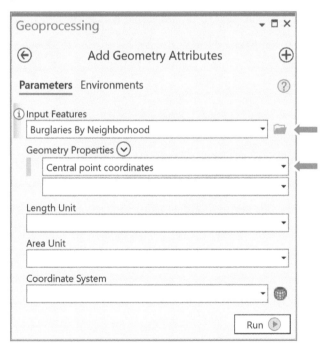

3.   **Open the Burglaries By Neighborhood attribute table.** The tool created INSIDE_X and INSIDE_Y central point coordinates, in state plane feet.

4. **Close the attribute table.**

Create a point layer
The Feature To Point tool creates a point layer of central points for polygons.

1. **Search for and open the** Feature To Point **tool.**

2. **Select Burglaries By Neighborhood as Input Features, type** BurglariesByNeighborhoodPoints **for Output Feature Class, select Inside, and run the tool.**

YOUR TURN
Symbolize BurglariesByNeighborhoodPoints with graduated symbols using Join_Count with five quantiles. Although not required here, you can change the name of the Join_Count field to **Burglaries** to make the feature layer more self-documenting. You now have a point layer that you can use with a choropleth map of some other attribute by neighborhood, such as population living in poverty, and you can publish both to ArcGIS Online.

Tutorial 4-7: Create a new table for a one-to-many join

Data in a database commonly has a code field that, by itself, is not self-documenting. For example, crime data may have as crime type only the FBI hierarchy code, 1 through 7. The meaning of these codes is 1 = Criminal Homicide, 2 = Forcible Rape, 3 = Robbery, 4 = Aggravated Assault, 5 = Burglary, 6 = Larceny-Theft, and 7 = Motor Vehicle Theft. To break the code, you need a code table with the seven codes in one field and the descriptions in a second field. In this tutorial, you'll create the code table, create a template for its fields, enter data into it, and join the table to crime data. This join is called one to many, because the same code table record—for example, the one for homicide—is joined to many crime incidents, all of them homicides.

You can use a field domain (see chapter 12) as an alternative approach for adding code values such as crime type to a table. This alternative is most relevant when the descriptive name of the code is itself the code (and there are no cryptic code values such as the FBI hierarchy). The code values become part of the table. For example, when responding police officers enter crime data using ArcGIS Collector, Collector includes a drop-down list with all crime types available for selection. This way, an officer doesn't need to type the information. A drop-down list improves data integrity because crime type codes are not typed but selected from the drop-down list. This feature eliminates the potential of misspelled words and guarantees that the user will retrieve all relevant records—for example, after a query of all crimes of a certain type.

Open the Tutorial 4-7 project

1. **Open Tutorial4-7.aprx from Chapter4\Tutorials, and save it as** Tutorial4-7YourName.aprx**.**

2. **Use the Pittsburgh bookmark.**

Create a table

1. **Search for and open the** Create Table **tool.**

2. **For table name, type** UCRHierarchyCode **and run the tool.**

YOUR TURN

Open the Fields design view of UCRHierarchyCode, and add two new fields: **Hierarchy,** with the Short data type, and **Crime,** with the Text data type and length **25**. Then open the table, and type seven records: Criminal Homicide, Forcible Rape, Robbery, Aggravated Assault, Burglary, Larceny–Theft, and Motor Vehicle Theft. Note that ArcGIS Pro automatically enters values for ObjectID, so you don't have to type them. Click Edit, and click Save when finished.

OBJECTID	Hierarchy	Crime
1	1	Criminal Homicide
2	2	Forcible Rape
3	3	Robbery
4	4	Aggravated Assault
5	5	Burglary
6	6	Larceny-Theft
7	7	Motor Vehicle Theft

Make a one-to-many join

You'll join your new code table to the map's crime layer to make crime type available for use.

1. **Note that the attribute table for PittsburghSeriousCrimesSummer2015 has Hierarchy as a field but not crime type.**

2. **Right-click PittsburghSeriousCrimesSummer2015, click Joins and Relates > Add Join, and perform the join based on the Hierarchy field.**

3. **View the PittsburghSeriousCrimesSummer2015 attribute table again to see the joined attributes.**

YOUR TURN

Symbolize PittsburghSeriousCrimesSummer2015 with unique values and Crime as the value field. Although you could improve symbolization of crimes by giving each crime a different symbol, you'll not do that work here. You can see that by linking the code table, the legend in the Contents pane is now labeled with crime types. When you finish, save and close your project.

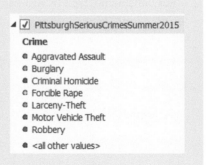

Assignments

This chapter has three assignments to complete that you can download from ArcGIS Online, at go.esri.com/GISTforPro2.8Data:

- **Assignment 4-1:** Investigate the spatial distribution of PhDs in Allegheny County.
- **Assignment 4-2:** Compare serious violent crime with poverty in Pittsburgh.
- **Assignment 4-3:** Query statistics for day and night burglaries and larcenies by month.

Spatial data

LEARNING GOALS

- Work with world map projections.
- Work with US map projections.
- Work with projected coordinate systems (PCS).
- Learn about vector data formats.
- Download US Census map layers and tabular data.
- Explore, download, and process data from ArcGIS® Living Atlas of the World.
- Explore sources of spatial data from government websites.
- Explore maps from a university's web service.

Introduction

The most important information that maps and GIS provide is location. Where is an object or event positioned? Is the object on a major street with good access? Is there a barrier to accessing it, such as a river? Is it in a certain area? Where are similar things positioned? What or who is near it? These are some of the questions that maps can answer.

Where you sit right now has two unique numbers, latitude and longitude coordinates that pinpoint your location precisely on the surface of the earth. In this chapter, you will learn some useful facts about latitude and longitude coordinates and their geographic coordinate system.

You will learn about map projections, making flat maps from the nearly spherical earth. You can project maps in at least 100 ways. Sometimes, the projection you choose makes a big difference, so this chapter provides some guidelines for choosing a projection.

For the world map in tutorial 5-1, you'll change the geographic coordinate system to a projected coordinate system by setting the map properties.

For the continental-level map in tutorial 5-2, you'll change the projected coordinate system to a different projected coordinate system by setting the map properties.

In tutorial 5-3, you'll set the projected coordinate system (state plane) for a local-level map by adding a first layer to the map and specifying the display units for the map. You'll add a second layer, which has geographic coordinates, and see how it's projected on the fly to the state plane coordinates. Finally, you'll change the geographic coordinate system for a state-level map to use a projected coordinate system by setting the map properties.

The key input of GIS is geospatial data—digital data that represents, and can be rendered into, maps on a computer. You may collect some original geospatial data yourself, but more likely, most if not all your map data will come from external organizations. Who collects aerial imagery, roads, population, or other map data that you will use? The answer is that international organizations and federal, state, or local governments collect the data and will provide it to you at little or no cost, as a download or increasingly as a service from the internet. ArcGIS Living Atlas of the World, available in ArcGIS Online, is one of the foremost sites for obtaining such data. You will explore Living Atlas and other US federal sites in this chapter.

Tutorial 5-1: World map projections

ArcGIS Pro has more than 5,200 projected coordinate systems (and almost 600 geographic coordinate systems) that use over 100 map projections from which you may choose. Typically, though, you will need relatively few projections for most purposes. Geographic coordinate systems use latitude and longitude coordinates for locations on the surface of the earth, whereas projected coordinate systems use a mathematical transformation from an ellipsoid (spheroid) or a sphere to a flat surface and rectangular coordinates. Geographic coordinates are angles calculated from the intersection of the prime meridian (which runs north and south through Greenwich, England) and the equator. Longitude, which measures east–west, ranges from 0 degrees to 180 degrees east and the same to the west. Latitude, which measures north–south, ranges from 0 degrees to 90 degrees north and the same to the south. Although you can view 2D maps in geographic coordinates on your flat computer screen, they are greatly distorted because their coordinates are from a sphere. In ArcGIS Pro, you can switch between coordinate systems and map projections on the fly. For world maps, you'll change the geographic coordinate system to a projected coordinate system by setting the map properties.

Open the Tutorial 5-1 project

This exercise shows you the distortions caused by displaying a map in geographic, latitude and longitude coordinates.

1. **Open Tutorial5-1.aprx from Chapter5\Tutorials, and save it as** Tutorial5-1YourName.aprx.

2. **Zoom to full extent.** The map has significant distortions. For example, the line running across the top and bottom of the map should be points, the North and South Poles. Also, Antarctica and Greenland are far larger proportionately than in reality. Although perhaps not evident, the Lower 48 states of the United States are squashed in the vertical dimension, as are Europe and Asia.

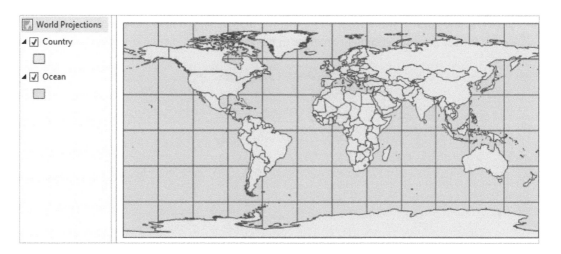

3. Examine a world map in geographic coordinates.

4. Place your pointer over the westernmost point of Africa, and read the coordinates on the bottom of the map window (appearing approximately at 17° W, 20° N). The network of lines on the map is called a graticule, and it has 30-degree intervals east–west and north–south.

Project the map on the fly to Hammer-Aitoff (world)

The projected coordinate system Hammer-Aitoff works well on a world map because it is an equal-area projection that preserves area. For example, if you want to map population densities (such as population per square mile), the densities will be correct.

1. In the Contents pane, right-click World Projections > Properties > Coordinate Systems. The current coordinate system is GCS WGS 1984.

2. Under XY Coordinate Systems Available, scroll down and click to expand Projected Coordinate System > World, and click Hammer-Aitoff (world).

3. **Click OK, and zoom to full extent.** Note that the map displayed here has coordinates projected on the fly that are not permanent changes to the feature class.

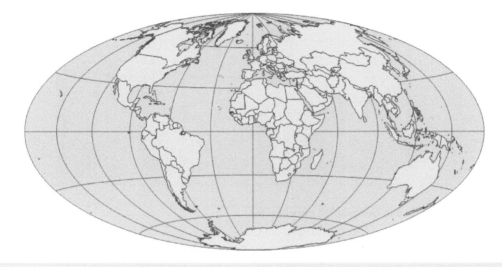

YOUR TURN

Repeat the steps of this exercise, but this time, select the Robinson (world) projection in the second step. The Robinson projection is most accurate at the mid-latitudes in both the Northern and Southern Hemispheres where most people live, and overall, it is appealing visually. This projected coordinate system minimizes some distortions but still has notable ones such as the lines at the top and bottom that in reality are points. As a rule, though, use the Robinson projection for the entire world unless you have a specific need (such as mapping population densities, in which case, use an equal-area projection). Save your project.

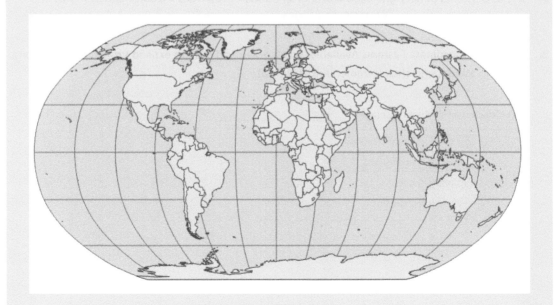

Tutorial 5-2: US map projections

Next, you will gain experience with projections commonly used for maps of the continental United States. You can get accurate areas or accurate shapes and angles, but not both, when making a projected coordinate system. As a rule, use a projection that provides accurate areas (at the price of some shape and direction distortion), such as the Albers equal area or cylindrical equal area projection. Albers equal area is the standard projection of both the US Geological Survey and the US Census Bureau for US maps. For a continental-level map, you'll change the projected coordinate system to a different projected coordinate system by setting the map properties.

Open the Tutorial 5-2 project

1. **Open Tutorial5-2.aprx from Chapter5\Tutorials, and save it as** Tutorial5-2YourName.aprx.

2. **Use the Contiguous Lower 48 States bookmark.** Initially, the map display is in projected coordinates using WGS 1984 Web Mercator (Auxiliary Sphere), which is the preferred PCS for maps to be published in ArcGIS Online.

Set projected coordinate systems for the United States

1. **In the Contents pane, right-click US Projections > Properties > Coordinate System, and scroll up and expand Projected Coordinate System > Continental > North America.**

2. **Click USA Contiguous Albers Equal Area Conic, and click OK.**

3.   **Zoom in to the Lower 48 states.** Your map is now using the Albers Equal Area projection.

YOUR TURN

Experiment by applying a few other projections to the US map such as North America Equidistant
Conic. As long as you stay in the correct group—Continental, North America—all the projections look similar. The conclusion is that the larger the part of the world that you need to project
(small-scale maps), the more distortion. There remains much distortion at the scale of a conti-
nent, but much less so than for the entire world. By the time you get to a part of a state, such as
Allegheny County, Pennsylvania, practically no distortion is left, as you will see next. Save your
project.

Tutorial 5-3: Projected coordinate systems

For medium- and large-scale maps, use localized PCSs, tuned for the study area, that have little or
minimal distortion. For this purpose, collections of PCSs are divided into zones. You must use a
reference map to determine which zone your study area is in and select that PCS for your map. In
these exercises, you'll set the PCS (state plane) for a local-level map by adding a first layer to the map,
and you'll specify the display units for the map. Then you'll add a second layer that has geographic
coordinates and see how that layer is projected on the fly to the state plane coordinates. Finally, you'll
change the geographic coordinate system for a state-level map to use a projected coordinate system
by setting the map properties.

Look up a zone in the state plane coordinate system

The state plane coordinate system is a set of coordinate systems dividing the 50 US states, Puerto Rico, and the US Virgin Islands, American Samoa, and Guam into 126 numbered zones, each composed of counties, and with its own finely tuned map projection. Used mostly by local government agencies such as counties, municipalities, and cities, the state plane coordinate system is for large-scale mapping in the United States. The US Coast and Geodetic Survey (now known as the National Geodetic Survey) developed this coordinate system in the 1930s to provide a common reference system for surveyors and mapmakers. Most states use NAD83, which stands for North American Datum of 1983, the datum used to describe the geodetic network in North America. That datum in turn was updated from the original North American Datum of 1927 when satellite geodesy and remote sensing technology became more precise and were made available for civilian applications. The first step in using the state plane coordinate system is to look up the correct zone for your area and consequently a specific PCS tailored to your study area.

1. **Start a web browser, and browse to Living Atlas:** livingatlas.arcgis.com.

2. **In the search box, type** USA State Plane Zones NAD83.

3. **Choose the feature layer map by Esri called USA State Plane Zones NAD83, and open it in Map Viewer.**

4. **Zoom to a familiar city or county, and click a zone to determine its zone abbreviation and number.**

5. **Close your web browser.**

Open the Tutorial 5-3 project

ArcGIS Pro offers two options for setting the map's coordinate system. One option is to allow the first map layer added to determine the coordinate system in each map according to its coordinate system. The other (default) option is to set a default coordinate system for all new maps in the project. In this exercise, you set the option to allow the first layer added to set the coordinate system for the project—in this case, State Plane Pennsylvania South.

1. **Open Tutorial5-3.aprx from Chapter5\Tutorials, and save it as** Tutorial5-3YourName.aprx. The project opens with a basemap and the default PCS, WGS 1984 Web Mercator (Auxiliary Sphere). You will change the option to have ArcGIS Pro set the coordinate system using the first layer added to the map—in this exercise, the state plane PCS.

Use state plane coordinates

1. On the Project tab, click Options > Map And Scene, expand Spatial Reference, and click Use
    Spatial Reference Of First Operational Layer if it is not checked. Click OK. Note that here is
    where you can set the map's PCS or GCS, using the Choose Spatial Reference option regardless
    of what layers you add to it. Next, you will add a layer with a state plane coordinate system PCS
    and then a layer using a GCS.

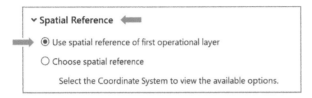

2. Click the back button, and on the Map tab, in the Layer group, click Add Data, browse to
    Chapter5.gdb, and add Municipalities. This layer for the municipalities of Allegheny County,
    Pennsylvania, uses zone 3702 PA South State Plane coordinates, and the map unit is US survey
    feet. Next, you must change the display units to feet.

3. In the Contents pane, right-click Allegheny County State Plane > Properties > General, select
    (or confirm) US Feet as the display units, and click OK. The x,y coordinates at the bottom of the
    screen now display as state plane feet.

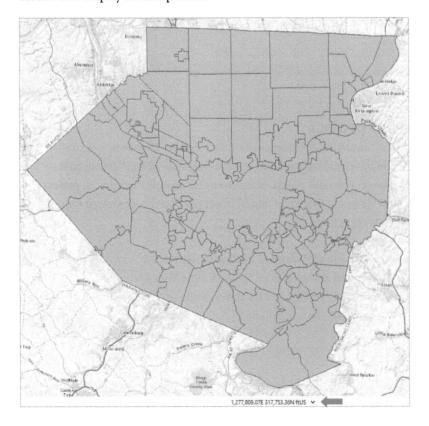

Add a layer using geographic coordinates

1. **Add Tracts from Chapter5.gdb.** This layer has geographic coordinates but is projected on the fly in the data frame to state plane coordinates because you added Municipalities first, which has state plane coordinates.

2. **Change the symbology of Municipalities to No Color, black outline width to** 1.5, **Tracts to white, outline color to gray 30%, and width to** 1 pt.

3. **Move Municipalities above Tracts in the Contents pane.**

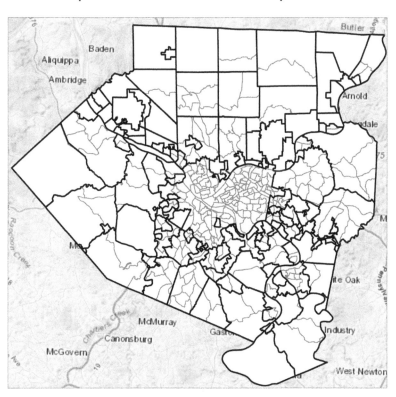

View a layer's spatial reference

1. **In the Contents pane, right-click Municipalities > Properties > Source, and expand Spatial Reference.** The spatial reference for this layer is NAD 1983 StatePlane Pennsylvania South FIPS 3702 Feet, which uses the map projection Lambert Conformal Conic.

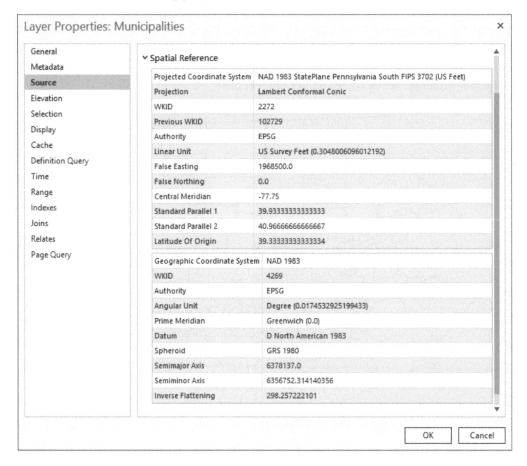

2. **Repeat step 1 to see the spatial reference for Tracts.** Its spatial reference is GCS North American 1983.

3. **Save your project.**

Use UTM coordinates

The US military developed the Universal Transverse Mercator (UTM) grid coordinate system in the late 1940s. It covers the world with 60 longitudinal zones defined by meridians (longitude lines) that are six degrees wide. ArcGIS has UTM zones available for the Northern and Southern Hemispheres. These zones, like state plane, are good for areas about the size of a state (or smaller).

Change a map's coordinate system to UTM

1. **Close the Allegheny County State Plane map, and open the California UTM map.** Note that the Counties layer has geographic coordinates.

2. **In the Contents pane, right-click California UTM > Properties > Coordinate Systems.**

3. **Expand Projected Coordinate System > UTM > NAD 1983, click NAD1983 UTM Zone 11N, and click OK.** The map coordinates are in decimal degrees. Next, you will change these coordinates to meters.

4. **In the Contents pane, right-click California UTM > Properties > General, choose Meters for Display Units, and click OK.**

5. **Use the California bookmark, and hover the pointer in the center of the state.** The x,y coordinates at the bottom of the display are UTM meters.

> ### YOUR TURN
>
> On the Project tab, click Options > Map and Scene > Spatial Reference. Select Choose Spatial Reference and browse to Projected Coordinate System > World > WGS 1984 Web Mercator (Auxiliary Sphere), and click OK. Insert a new map, add Counties, and then view the coordinate system of the map. It's Web Mercator, the preferred format for ArcGIS Online. Save your project.

Tutorial 5-4: Vector data formats

This tutorial reviews file formats commonly found for vector spatial data, in addition to file geodatabases covered in chapter 4. Included are Esri shapefiles, x,y data, and Google KML files.

Examine a shapefile

Many spatial data suppliers use the shapefile data format, an Esri legacy format, for vector map layers because it is so simple. A shapefile consists of at least three files with the following extensions:.shp,.dbf, and.shx. Each file uses the shapefile's name but with a different extension (for example, Cities.shp, Cities.dbf, and Cities.shx). The SHP file stores the geometry of the features, the DBF file stores the attribute table, and the SHX file stores an index of the spatial geometry to speed up processing. Next, you will examine a census tract shapefile in more detail.

1. **Open a File Explorer window, and browse to Chapter5\Data\.** CouncilDistricts is a shapefile for New York City Council Districts provided by the city's Planning Department. Shapefiles have three to seven associated files, including the three mentioned previously (.shp,.shx, and.dbf) and also a PRJ file that contains the map layer's coordinate system.

Name ^	Type	Size
CouncilDistricts.dbf	DBF File	4 KB
CouncilDistricts.prj	PRJ File	1 KB
CouncilDistricts.shp	SHP File	654 KB
CouncilDistricts.shp	XML Document	8 KB
CouncilDistricts.shx	SHX File	1 KB

2. **Close the File Explorer window.**

Import a shapefile into a file geodatabase and add to a map

You will use a conversion tool to import a shapefile into a file geodatabase, where it becomes a feature class.

1. **Open Tutorial5-4.aprx from Chapter5\Tutorials, and save it as** Tutorial5-4YourName.aprx. The project opens with a map zoomed to New York City with a PCS of NAD 1983 StatePlane New York Long Isl FIPS 3104 (US Feet).

2. **On the Analysis tab, click the Tools button.**

3. **Search for and open the** Feature Class to Feature Class **tool.**

4. **For Input Features, browse to the Data folder and add CouncilDistricts.shp. For Output Feature Class, type CouncilDistricts, and run the tool.**

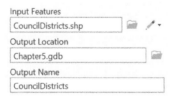

The shapefile is now a polygon feature class in Chapter5.gdb and is automatically added to the Contents pane. The council districts nicely overlay on the basemap. Note the state plane coordinates at the bottom of the map.

Add x,y data

GPS units and many databases provide spatial coordinates as x,y coordinates. You can add these to a map as XY Event files. In this exercise, you will use x- and y-values from a table to create a point feature class of public libraries in New York City.

1.   **On the Map tab, click Add Data, browse to Chapter5\Data\, and add Libraries.dbf.**

2. **In the Contents pane, right-click Libraries > Open, and scroll to the right to see columns XCOORD and YCOORD.** These coordinates are in state plane format for New York City. The x,y locations are often obtained using GPS units and are in longitude and latitude format. In these cases, x is the longitude and y is the latitude, and the spatial reference is WGS84 or a similar geographic coordinate system.

OID	OBJECTID	BOROUGH	FACILITY_N	FACILITY_A	ZIP	FACILITY_T	COUNCIL	XCOORD	YCOORD	Factype_tx	Factype_1
0	2198	MN	NEW AMSTERDAM L...	9 Murray St	10007	1401	1	982074	199262	1401	Public Library - Branch
1	2199	MN	HUDSON PARK LIBR...	66 Leroy St	10014	1401	3	982779	205207	1401	Public Library - Branch
2	2200	MN	JEFFERSON MARKET...	425 6 Ave	10011	1401	3	984425	206857	1401	Public Library - Branch
3	2201	MN	CHATHAM SQUARE...	33 E Broadway	10002	1401	1	985223	199159	1401	Public Library - Branch
4	2202	MN	HAMILTON FISH PA...	415 E Houston St	10002	1401	2	989910	201380	1401	Public Library - Branch
5	2203	MN	OTTENDORFER LIBR...	135 2 Ave	10003	1401	2	987632	204884	1401	Public Library - Branch
6	2216	MN	SEWARD PARK LIBRA...	192 E Broadway	10002	1401	1	987425	199572	1401	Public Library - Branch

3. **Close the table.**

4. **In the Contents pane, right-click Libraries > Display XY Data.**

5. **In the Display XY Data pane, type** Libraries **for Output Feature Class in Chapter5.gdb, verify XCOORD is the X field and YCOORD is the Y field, click Current Map [Vector Data Formats] for Coordinate System, and run the tool.** The spatial reference automatically sets itself to NAD_1983_StatePlane_New_York_Long_Island_FIPS_3104_Feet.

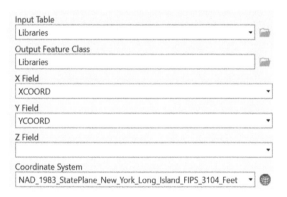

Chapter 5: Spatial data

GIS Tutorial for ArcGIS Pro 2.8

161

Part

2

Chapter

5

Tutorial

4

The libraries appear as points on the map.

6. Remove the Libraries layer and the Libraries table from the Contents pane, and save your project.

Export to a Google KML file

Google KML (Keyhole Markup Language) is the file format used to display geographic data in many mapping applications. KML has become an international standard maintained by the Open Geospatial Consortium (OGC). In this exercise, you will export the New York City Council Districts to Google KML format.

1. **Search for and open the** Layer To KML **tool.**

2. **In the pane, for Layer, select CouncilDistricts, and for Output File, select CouncilDistricts.kmz.**

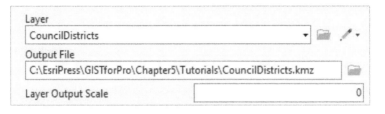

3. **Run the tool.**

4. **Save your project.** Although you won't do it here, you can now open the.kmz file, the compressed version of a KML file, in an application such as Google Earth.

Tutorial 5-5: US Census map layers and data tables

A large and ever-expanding collection of spatial data—including vector and raster map layers and data tables with geocodes—is available for free download from the internet or for direct use as map layers in projects. The advantage of downloading spatial data is that you can modify it. In this tutorial, you will learn about and use data from the US Census Bureau, a major spatial data supplier.

From the US Census Bureau's website, you can download census-related map layers with unique identifiers for polygons such as census tract IDs but not census attribute data such as population, age, sex, race, income, and so on. Because there are thousands of census variables, it is impractical to provide them all in map layer attribute tables. The map layers would be excessively large and cumbersome files. Instead, the Census Bureau provides websites in which you can build custom collections of variables for your needs. The Census Bureau continually updates these websites, so you might find that some popular collections of data are already available and that the website's menus may vary.

In this tutorial, you will download selected census demographic data separately, perform data preparation steps, and finally join the data to the corresponding polygon boundary map layer. First, you will download Topologically Integrated Geographic Encoding and Referencing (TIGER) shapefiles for tracts and county subdivisions of Hennepin County, Minnesota, where the city of Minneapolis is located.

Download census TIGER files

1. **In a web browser, go to census.gov/cgi-bin/geo/shapefiles/index.php.** There are alternative methods for downloading TIGER files such as file geodatabases. This is a direct link to individual shapefiles.

2. **Select 2010 for year and Census Tracts for layer type, and click Submit.** Although there are newer dates, select 2010 for layers from the decennial census of 2010.

3. Under Census Tract (2010), select Minnesota; under Select a County, select Hennepin County; and click Download.

4. Save the compressed file and extract all files associated with the shapefile to Chapter5\ Tutorials\Downloads\Census.

YOUR TURN

Download and extract both the 2010 County Subdivisions and the 2010 Water (Area Hydrography) features for Hennepin County, Minnesota, to the Downloads\Census folder.

Download census tabular data

The Census Bureau provides detailed data on education, income, transportation, and other subjects through its American Community Survey (ACS) tables. In the 2000 and previous censuses, such data was collected in the census long form randomly from one out of six households and was called Summary File 3 (SF3) data. Now, the Census Bureau collects ACS data monthly—approximately 3 million housing units receive a survey like the old long-form questionnaire. Annual, three-year, and five-year estimates produced from ACS samples are available. In this exercise, you will download commuting-to-work ACS data at the tract level for Hennepin County, Minnesota. The city of Minneapolis is ranked high in the nation for the quality of its transportation infrastructure, including bike lanes, according to several groups and publications (including the League of American Bicyclists and Bicycling). The census is a good starting point to analyze means of transportation to work by male, female, and overall percentages. You will first select the geography (all tracts in Hennepin County) and the topic (commuting characteristics by sex).

1. In a web browser, go to data.census.gov, and click Advanced Search.

2. Click Geography > Tract > Minnesota > Hennepin County, Minnesota, and click All Census Tracts Within Hennepin County, Minnesota. This will add all tracts in this county to a selection.

3. Under Find a Filter, type Commuting and click Search.

4. Check the box for Commuting, and click Search. A list of related tables appears.

5. Click the first table, Commuting Characteristics by Sex (table S0801). A preview of the table appears.

6. Click Download, and click the check box for Commuting Characteristics By Sex.

7. **Click Download Selected, and modify the selected boxes so that the only one selected is the box for 2015.** Data is constantly added to the US Census Bureau website so there may be newer data sets for ACS five-year estimates that you could use for updated studies. This tutorial uses data from the 2010–2015 American Community Survey and will work nicely with the 2010 tracts.

8. **Click the Download button, wait while the CSV file is prepared, download the file, and extract it to the Downloads\Census folder.** Note that the table will be named with the date you download it.

Process tabular data in Microsoft Excel

The census data that you downloaded in the previous steps needs some cleaning up using software such as Microsoft Excel before you use the data in GIS. You will delete all columns except those needed for joining and the estimate of percentages of males and females who bicycle to work.

1. **Browse to Chapter5\Tutorials\Downloads\Census, and double-click the comma-separated file named ACSST5Y2015.S0801_data_with_overlays_date.csv.** The file opens in Excel.

2. **Delete the first row.**

3. **Delete all fields except the fields listed below. Hint: Select all fields, hold down the Control key to deselect the fields listed below, right-click, and select delete.**
    *   **Id. Column A**
    *   **Total!!Estimate!!MEANS OF TRANSPORTATION TO WORK!!Bicycle Column W**
    *   **Male!!Estimate!!MEANS OF TRANSPORTATION TO WORK!!Bicycle. Column EG**
    *   **Female!!Estimate!!MEANS OF TRANSPORTATION TO WORK!!Bicycle. Column IQ**

4. **Rename these fields:** GEOID, TOTAL_BIKE, MALE_BIKE, **and** FEMALE_BIKE.

5. **Highlight column A.**

6. **On the Home tab, in the Editing group, click Find & Select, and choose Find.**

7. **In the Find and Replace dialog box, click the Replace tab. For the Find What field, type** 1400000US, **leave the Replace With field blank, and click Replace All.** This converts the census tract GEOID field to a number that includes just the tract numbers. The field type is not yet numeric so you will need to change this.

8. **Highlight fields TOTAL_BIKE, MALE_BIKE, and FEMALE_BIKE, right-click, and click Format Cells.**

9. **In the Format Cells dialog box, under Category, choose Number, one decimal place, and click OK.**

10. **Highlight the GEOID field, right-click, and click Format Cells, and choose Number and zero decimal places.**

11. **Save the file as** BikeWorkData.csv.

	A	B	C	D
1	GEOID	TOTAL_BIKE	MALE_BIKE	FEMALE_BIKE
2	27053000101	1.0	1.9	0.0
3	27053000102	0.5	0.0	1.0
4	27053000300	0.4	0.9	0.0
5	27053000601	1.9	2.9	0.8
6	27053000603	0.6	1.2	0.0

Add and clean data in ArcGIS Pro

1. **Open Tutorial5-5.aprx from Chapter5\Tutorials, and save it as** Tutorial5-5YourName.aprx.

2. **Import your downloaded county subdivisions (tl_2010_27053_cousub10.shp) into Chapter5.gdb as** HennepinCountySubdivisions, **census tracts (tl_2010_27053_tract10.shp) as** HennepinTracts, **water features (tl_2010_27053_areawater.shp) as** HennepinWater, **and your comma-delimited file (BikeWorkData.csv) as** BikeWorkData. The features and table will automatically be added to Contents.

3. **Open the table for BikeWorkData, and sort ascending for the field GEOID.** Note that GEOID is a numeric field.

OBJECTID *	GEOID ▲	TOTAL_BIKE	MALE_BIKE	FEMALE_BIKE
1	27053000101	1	1.9	0
2	27053000102	0.5	0	1
3	27053000300	0.4	0.9	0
4	27053000601	1.9	2.9	0.8
5	27053000603	0.6	1.2	0
6	27053001100	1.4	2.5	0

4. **Close the BikeWorkData table, and open the attribute table for HennepinTracts.** The candidate field for joins is GEOID10 and is a text field.

5. **Open Fields View from the Options button (upper right) of the table, and create a numeric field (whose field type is double) named** GEOIDNUM.

6. **Save and close Fields View.**

7. Use the Calculate Field tool to set GEOIDNUM equal to GEOID10. This step is necessary to join tables in which the join field must be the same type.

8. Close the HennepinTracts table.

Join data and create a choropleth map

1. Join the BikeWorkData table to HennepinTracts using fields GEOIDNUM (Input Join Field) and GEOID (Join Table Field).

2. Export the joined table and features as HennepinTractsBikeWork to Chapter5.gdb, and copy and paste the layer so there are two copies.

3. Rename one layer % Male bicyclists and the other layer % Female bicyclists.

4. Create a choropleth map using the field MALE_BIKE, the color scheme Purples (5 classes), and upper values 1.5%, 3%, 6%, 12%, and 31.2%.

5. Create the same map using the field FEMALE_Bike, the color scheme Orange (5 classes), and upper values 1.5%, 3%, 6%, 12%, and 19.3%.

6. Remove the original HennepinTracts layer and BikeWorkData table.

7. Move HennepinCountySubdivisions to the top of the Contents pane, change its display to No Color, and set a black outline width of 1.5.

8. Label this layer using Name10 with a white halo.

9. Move HennepinWater below HennepinCountySubdivisions, and change its symbology to Water (area).

10. **Zoom to Minneapolis, and turn the layers on and off to see the difference in percentage of males versus females who bicycle to work, and then save your project.** Do males or females bicycle to work in different parts of Minneapolis at a higher percentage?

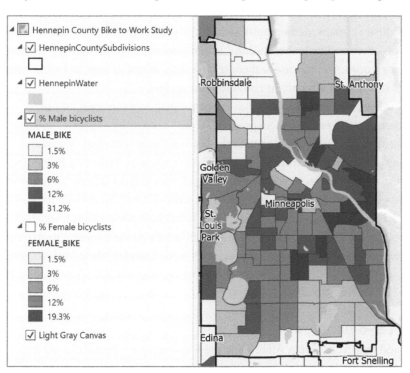

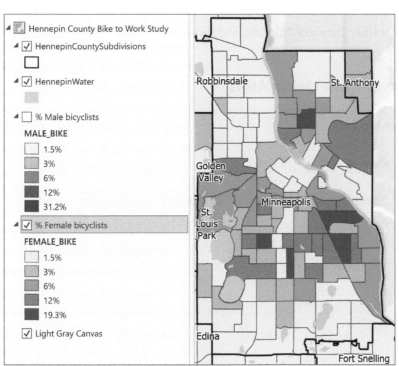

Tutorial 5-6: Download data from Living Atlas

You can access and quickly use content from ArcGIS Living Atlas of the World. Living Atlas is one of the foremost collections of geospatial data around the world. Data includes maps and data on thousands of topics in the form of imagery, basemaps, and features that can be added directly or saved locally. Examples include historic maps, demographics, and lifestyles for the United States and many other countries, landscapes, oceans, Earth observations, urban systems, transportation, boundaries and places, and StoryMaps stories. You can use and contribute data, maps, or apps to Living Atlas. Data is frequently added, so see livingatlas.arcgis.com often for updates.

In this exercise, you will search for and add a land-use raster map and extract it for only Hennepin County, Minnesota. Whereas vector maps are discrete—consisting of points, lines, and polygons connecting coordinates—raster maps such as the land-use map represent continuous phenomena and use many of the same file formats as images on computers, including JPEG (.jpg) and TIFF (.tif) formats. All raster maps are rectangular, consisting of rows and columns of cells known as pixels. Each pixel has associated x,y coordinates and a z-value attribute, such as altitude for elevation or some other property. Raster maps do not store each pixel's location explicitly but rather store data such as the coordinates of the northwest corner of the map, cell size (assuming square pixels), and the number of rows and columns, from which a computer algorithm can calculate the coordinates of any cell.

Open the Tutorial 5-6 project

In ArcGIS Pro, you can add data from Living Atlas using the Catalog pane > Portal (Living Atlas) or using the Add Data button.

> Open Tutorial5-6.aprx from Chapter5\Tutorials, and save it as Tutorial5-6YourName.aprx.

Add a land-use layer from Living Atlas

1. On the Map tab, click Add Data.

2. Click Portal > Living Atlas.

3. In the Search Portal: Living Atlas window, type NLCD, and press Enter. The list of available layers may vary from the figure.

4. **Click USA NLCD Land Cover, and click OK.** The raster layer loads. A land-use map for Hennepin County, Minnesota, with an updated legend shows areas of land use.

Extract raster features for Hennepin County

A key aspect of raster maps is that they are large files. So, although you may store some important basemap raster files on your computer, these maps are perhaps best obtained as map services available for display on your computer but stored elsewhere such as Living Atlas. If you want to extract a subset of the land use for one county and store the raster file on your local computer, you must use the Extract By Mask tool to do so.

1. **Search for and open the** Extract By Mask **tool.**

2. **Select USA NLCD Land Cover for Input Raster and HennepinCounty as the feature mask data, save the output raster as** HennepinCountyLandUse **to Chapter5.gdb, and run the tool.**

3. **Remove the original land-use layer, zoom to the Hennepin County land-use layer, and save your project.** Notice the difference in Hennepin County with developed areas on the eastern side of the county and the forest, pasture/hay, and cultivated crops in the western part of the county.

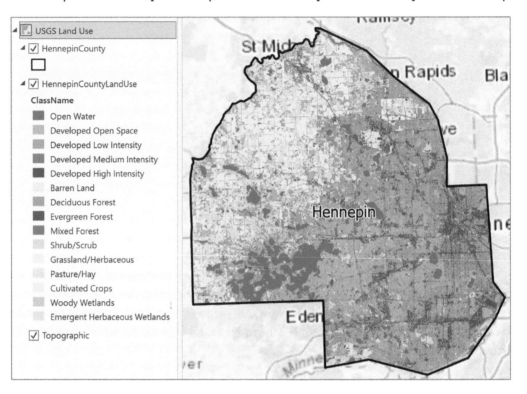

YOUR TURN

Search for topics (for example, health or housing) or geographic areas (for example, China). Add and explore the contents of various Living Atlas layers.

Tutorial 5-7: Sources of data from government websites

In this tutorial, you will download data from two US government websites: USGS's National Map Viewer and Data.gov. Agencies that contribute geospatial data include the US Department of Agriculture (USDA), Department of Commerce (DOC), National Oceanic and Atmospheric Administration (NOAA), US Census Bureau (Census), Department of the Interior (DOI), US Geological Survey (USGS), Environmental Protection Agency (EPA), and National Aeronautics and Space Administration (NASA). Many other agencies provide their data on federal websites such as Data.gov.

Download elevation contours from USGS

Walkability and bicycling studies often require knowledge of the topography of a city, and elevation contours are needed. You can get these contours from local GIS departments or USGS.

1. In a web browser, go to apps.nationalmap.gov/downloader/#/.

2. Search for Minneapolis MN in the map's search box. You can also manually zoom to the area.

3. Select Elevation Products (3DEP), and select only the check box for Contours.

4. Under File Formats, click Shapefile, and click Search Products.

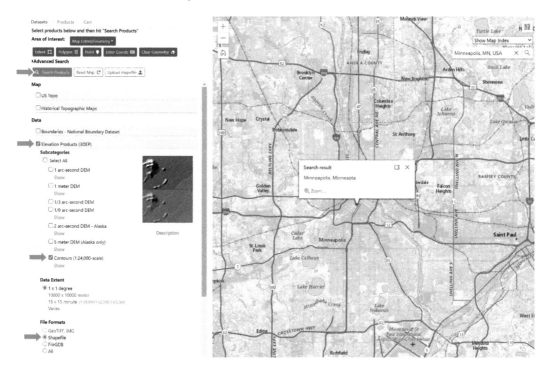

5. With multiple extents possibly available, choose the file for Saint Paul W, Minnesota. Click the Add This Item To The Cart button.

6. Click the Cart tab and click Download ZIP. Wait while the file downloads.

7. Open the compressed downloaded file, and extract its contents to Chapter5\Tutorials\ Downloads\USGS.

8. Open Tutorial5-7.aprx from Chapter5\Tutorials, and save it as Tutorial5-7YourName.aprx.

9. On the Map tab, click Add Data, and browse to Chapter5\Tutorials\Downloads\USGS and the subfolder created.

10. **Add the file, Elev_Contour to the map; in Contents, drag it below Minneapolis; and zoom to the Minneapolis layer.** The map zooms to Minneapolis and contours of the southern part of the city, where you can see areas of the city that are flat. In chapter 7, you will learn how to clip these features to the city's boundary.

11. **Symbolize the contours as ground features, and save your project.**

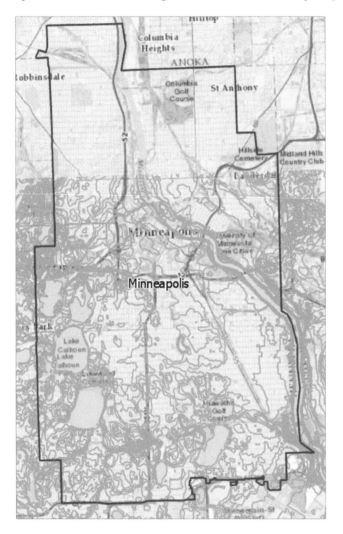

Download 311 data from Data.gov

Data.gov is the location of the US government's open data, where you find many useful datasets to use in your GIS projects. It includes a wide range of topics ranging from climate and energy to law and finance, to name a few. It also includes links to resources such as cities and other local governments, and you can search by specific organization. In this exercise, you learn how to search for local government (county or city) 311 (nonurgent reporting or service request) data that you can use in GIS.

1. **In a web browser, go to catalog.data.gov/dataset, and in the search box, type 311 and press Enter.** The type of data (for example, shapefile, Excel, KMZ) is listed below each dataset along with the source (for example, city, county, federal, university, and so on).

2. **Under Topics, click Local Governments.** This selection restricts the search to counties and cities.

3. **Click 311 Data for Allegheny County/City of Pittsburgh/Western PA Regional Data Center.** This is a dataset of 311 service calls for the City of Pittsburgh that were collected from phone calls, tweets, emails, a form on the city website, and through the 311 mobile app.

4. **Click the Download button next to 311 Data CSV, save the file to Chapter5\Tutorials\ Downloads\DataGov, and rename the file** Pittsburgh311Data.csv.

5. **Close the USGS Contours map, and open the Pittsburgh 311 Data map.** The map includes the municipalities of Allegheny County, Pennsylvania, with Pittsburgh in its center. There are over 400,000 records so you will want to query by request type as you import it.

6. **Use the Table To Table tool, create the expression REQUEST_TYPE is equal to Food Safety Complaints, and type the Output Name text** FoodSafetyComplaints **to import the Pittsburgh311Data.csv file into the Chapter5 geodatabase.**

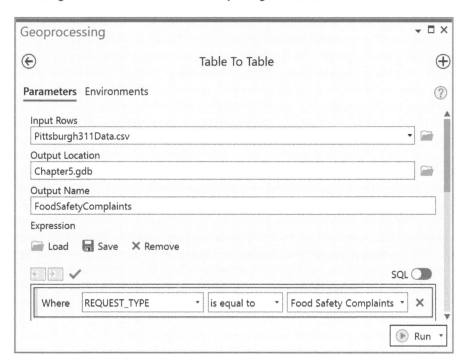

7.   **Run the tool, and close it after it finishes.**

8.   **Search for the tool Make XY Event Layer.**

9.   **For XY Table, select FoodSafetyComplaints. For X Field, select X. For Y Field, select Y. For Layer Name, type FoodSafetyComplaints_Layer, and for Spatial Reference, select GCS_WGS_1984. Run the tool.**

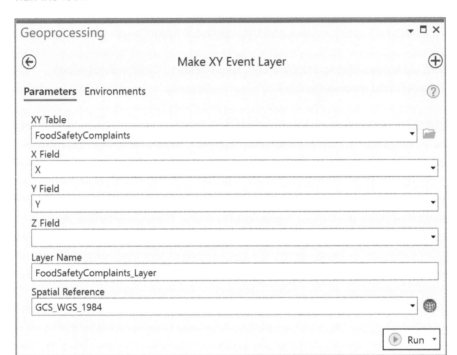

The map features of locations for food safety complaints are added to your map in the correct geographic location.

YOUR TURN

Search for and explore public datasets of interest from agencies such as USDA, NOAA, US Census Bureau, USGS, or NASA. Many other agencies, including local governments, provide their data at Data.gov.

Tutorial 5-8: Maps from a university web service

Federal, state, and local governments are great sources of geospatial data. Universities and libraries are also clearinghouses for data. They often use ArcGIS map services that allow you to connect to servers to directly access data. In this exercise, you will connect to Penn State University's Pennsylvania Spatial Data Access (PASDA) ArcGIS map service and add imagery data.

Open the Tutorial 5-8 project

1. **Open Tutorial5-8.aprx from Chapter5\Tutorials, and save it as** Tutorial5-8YourName.aprx. The project opens with Pittsburgh parks displayed as semitransparent features and a yellow outline.

Connect to ArcGIS server map service

1. **On the Insert tab, in the Project group, click Connections > New ArcGIS Server.**

2. **In the Add GIS Server User Connection window, for Server URL, type** https://imagery.pasda.psu.edu/arcgis/rest/services, **and click OK.** All the data and services provided by PASDA are free and open to the public without a password.

3. **On the Map tab, in the Layer group, click the Add Data button. From the Add Data window, under Project > Servers, click Services on imagery.pasda.psu.edu.ags, and click Open.**

4. **Double-click pasda, click AlleghenyCountyUrbanTreeCanopy2010, and click OK.**

5. **In the Contents pane, expand the AlleghenyCountyUrbanTreeCanopy2010 layer, and if needed, select the check box for AlleghenyCountyUrbanTreeCanopy2010.**

6. **Use the Point State Park bookmark.** The details of the tree canopy in Pittsburgh's Point State Park (where the three rivers meet) and parks along the riverfront are visible in the raster map.

YOUR TURN

Explore other map layers to add from PASDA's web service. Try to find another university or library that provides ArcGIS web services. Save your project.

Assignments

This chapter has two assignments to complete that you can download from ArcGIS Online, at go.esri. com/GISTforPro2.8Data:

- **Assignment 5-1**: Compare heating fuel types by US counties.
- **Assignment 5-2**: Study solar-heated homes by neighborhood.

6

Geoprocessing

LEARNING GOALS

- Dissolve block group polygons to create neighborhoods and fire battalions and divisions.
- Extract a neighborhood using attributes to form a study area.
- Extract features from other map layers using the study area.
- Merge water features to create a single water map.
- Append separate fire and police station layers to one layer.
- Intersect streets and fire companies to assign street segments to fire companies.
- Union neighborhood and land-use boundaries to create detailed polygons of neighborhood land-use characteristics.
- Apportion data between two polygon map layers whose boundaries do not align.

Introduction

Geoprocessing is a framework and set of tools for processing geographic data. Generally, you must use geoprocessing tools to build study areas in a GIS. In this chapter, you will learn how to extract a subset of spatial features from a map using attribute or spatial queries; aggregate polygons into larger polygons; append layers to form a single layer; and use Intersect, Union, and Tabulate Intersection tools to combine features and attribute tables for processing. In this chapter, for example, you'll process and prepare map layers for emergency management officials in New York City's Manhattan borough and one of its neighborhoods, the Upper West Side.

Tutorial 6-1: Dissolve features to create neighborhoods and fire divisions and battalions

Suppose emergency response planners want to know the number of housing structures or units by neighborhood for planning purposes. New York City neighborhoods are composed of block groups that include housing data. In this tutorial, you'll create neighborhood boundaries in the Manhattan borough by dissolving block groups with common neighborhood names. For each neighborhood, dissolving retains the neighborhood's outer boundary lines but removes interior lines. The Dissolve tool can aggregate block group attributes to the neighborhood level, using statistics such as sum, mean, and count. In this case, you'll sum the number of structures in housing units by neighborhood.

Open the Tutorial 6-1 project

1. **Open Tutorial6-1.aprx from the Chapter6\Tutorials folder, and save the project as** Tutorial6-1YourName.aprx.

2. **Use the Manhattan bookmark.** The map contains Manhattan block groups with housing units and structures.

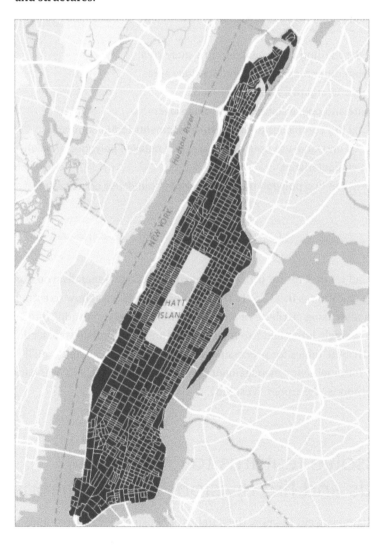

Examine the Dissolve field and other attributes

The Dissolve tool needs data—a Dissolve field, for combining block groups into a neighborhood. In this case, Name is the Dissolve field, and that field contains the neighborhood name to which each block group belongs.

1. **Open the ManhattanBlockGroups attribute table, and sort the Name field in ascending order.** As you scroll, you can see the block group records that make up each neighborhood. The figure shows some of the block groups for the Battery Park City–Lower Manhattan neighborhood and the associated GEOID.

OBJECTID	Shape	STATEFP	COUNTYFP	GEOID	Name ▲
18	Polygon	36	061	360610013002	BatteryParkCity-LowerManhattan
21	Polygon	36	061	360610015011	BatteryParkCity-LowerManhattan
22	Polygon	36	061	360610015021	BatteryParkCity-LowerManhattan
107	Polygon	36	061	360610007001	BatteryParkCity-LowerManhattan
111	Polygon	36	061	360610009001	BatteryParkCity-LowerManhattan
114	Polygon	36	061	360610013001	BatteryParkCity-LowerManhattan
638	Polygon	36	061	360610015013	BatteryParkCity-LowerManhattan
816	Polygon	36	061	360610317043	BatteryParkCity-LowerManhattan
817	Polygon	36	061	360610317044	BatteryParkCity-LowerManhattan

2. **Scroll to the right, and examine the remaining attributes.** Values are an estimate of the number of housing units in structures of different sizes, along with the total number of housing units in each block group. For example, the field TOT1_Detached is a one-family house detached from any other house. Mobile is the total number of mobile homes in a block group. TOT20_49 includes the total number of housing units for each block group in structures containing 20 to 49 units (that is, larger apartment or condo buildings). TOT50 is the total number of housing units in structures containing 50 or more units (mainly high-rise apartment and condo buildings). When dissolving block groups, you'll sum these housing unit values by neighborhoods, which will indicate housing density.

3. **Close the table.**

Dissolve block groups to create neighborhoods

1. **On the Analysis tab, in the Tools group, open the Dissolve tool.**

2. **In the Dissolve pane, under Input Features, select ManhattanBlockGroups.**

3. **Type** ManhattanNeighborhoods **as the output feature class in Chapter6.gdb., and select Name as the Dissolve field.**

4. **Click Add Many from the Statistics Field(s) option, scroll down, check the check box next to each of the TOT fields, and click Add.**

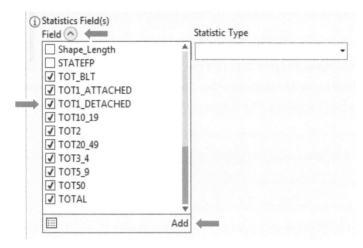

5. **Confirm Sum is the Statistic Type for each statistics field and run the tool.**

6. **In the Contents pane, open the ManhattanNeighborhoods attribute table, and click Sort Descending on the Sum_TOT50 field.** For example, on the Upper West Side, there are 41,240 housing units in structures having 50 or more units.

Name	SUM_TOT1_	SUM_TOT1_DETACHED	SUM_TOT10_19	SUM_TOT2	SUM_TOT20_49	SUM_TOT3_4	SUM_TOT5_9	SUM_TOT50 ▾
Upper West Side	543	500	8810	481	15806	1499	6093	41240
Lenox Hill - Roosevelt Island	246	222	5080	204	8785	483	2025	34976
Lincoln Square	211	207	2554	275	4480	244	1609	31340
Turtle Bay - East Midtown	70	268	1664	176	4382	268	750	29618
Yorkville	222	391	9061	85	9510	570	1543	27618

7. **Close the tables and map, and open the Fire Companies Battalions and Divisions map.**

YOUR TURN

Other useful map layers for city officials and emergency response teams show populations per square mile in fire battalions and divisions. The New York City Fire Department, like most fire departments around the country, is organized in a quasi-military fashion with companies, battalions, and divisions. Here, you have fire companies but need fire battalions and divisions. Use the Dissolve tool and attribute FireBN to dissolve fire company polygons to create a feature class named ManhattanFireBattalions in Chapter6.gdb. Use the Sum option and attributes Pop2010 and SQ_MI to sum the population and square miles for each battalion. Repeat Dissolve to create a feature class named ManhattanFireDivisions using field FireDiv as the Dissolve field. Symbolize ManhattanFireDivisions as No Color and a thick red outline and ManhattanFireBattalions as graduated colors (Blues, 5 classes) and normalized using Sum_Pop2010/Sum_SQ_MI. Remove ManhattanFireCompanies, and label the fire battalions using the battalion number, FireBN. The fire battalion for the Upper West Side neighborhood is 11. Is this one of the most densely populated fire battalion areas? Save your project.

Tutorial 6-2: Extract and clip features for a study area

This tutorial is a workflow for creating a study region from map layers that have more features than needed. Suppose you want to study housing units and streets for just one neighborhood. First, you will create the study area neighborhood by selecting the neighborhood using the attribute table and creating a single polygon for that study area. Second, you will use the new polygon and select by location to create features of block groups in the study area only. Third, you will use Clip, a geoprocessing tool, to clip streets to the study area.

Open the Tutorial 6-2 project

In this exercise, you will create a study area that includes a polygon feature of the Upper West Side neighborhood and block groups and streets for that neighborhood only. Such a study area is important when you work in geographic areas, such as New York City, that have many streets and block groups.

1. **Open Tutorial6-2.aprx from the Chapter6\Tutorials folder, and save the project as** Tutorial6-2YourName.aprx. The map contains New York City neighborhoods, block groups, and Manhattan streets.

2. **Use the Upper West Side bookmark.**

Use Select By Attributes to create a study area

1. On the Map tab, in the Selection group, click the Select By Attributes button.

2. Under Input Rows, select New York City Neighborhoods. Click New Expression, and in the Where field, select Name, Is Equal To, and Upper West Side. Run the tool.

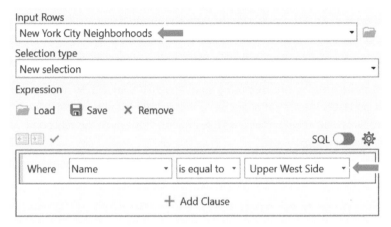

The result shows the Upper West Side neighborhood selected. Although you don't have to export the selected feature to select by location in the next section (Select By Location will use the selected neighborhood to select the block groups), you will export the feature to display and symbolize the Upper West Side neighborhood on a map.

3. In the Contents pane, right-click New York City Neighborhoods > Data > Export Features, type UpperWestSide as the output feature class in Chapter6.gdb, and run the tool.

4. Remove the original New York City Neighborhoods layer, and zoom to the Upper West Side.

Use Select Layer By Location to create study area block groups

In the next steps, you will use Select Layer By Location to select Manhattan block groups of the Upper West Side neighborhood. After creating a selection set of the block groups, you will create a polygon feature class from the selected features. Because block groups are not contiguous with the neighborhood boundary, there are a few ways to create this selection. Here, you will use the block group features whose centers are in the Upper West Side neighborhood and manually select the remaining block groups that are partially inside the neighborhood.

1. **On the Map tab, click the Select Layer By Location button. For Input Features, select New York City Block Groups. For Relationship, type** Have their center in**. For Selecting Features, select UpperWestSide. Run the tool.**

Input Features ⌄
New York City Block Groups ⬅

Relationship
Have their center in ⬅

Selecting Features
UpperWestSide ⬅

Search Distance
| | Feet |

Selection type
New selection

☐ Invert spatial relationship

2. **On the Map tab, click the Select button, press the Shift key, and select the remaining block groups on the west (left) side of the neighborhood.**

3. **Export the selected features as** UpperWestSideBlockGroups **to Chapter6.gdb.**

4.   Remove the original New York City Block Groups layer, and move the UpperWestSide layer to the top of the Contents pane.

YOUR TURN

Use Select By Location to select Manhattan streets that intersect the UpperWestSide polygon. Notice that many of the streets "dangle," or extend beyond the polygon boundary. These selected streets, each a full block long, are needed for geocoding address data to points, as explained in chapter 8. Save these as **UpperWestSideStreetsForGeocoding**, and clear the selected features.

Clip streets

Next, you will use the Clip tool to cleanly create street segments using the Upper West Side polygon. Clipping Manhattan streets with the Upper West Side boundary cuts off dangling portions of streets for display purposes in maps.

1.   Turn off the UpperWestSideStreetsForGeocoding and UpperWestSideBlockGroups layers, if necessary.

2.   On the Analysis tab, in the Tools gallery, open the Clip tool.

3.   Complete the parameters. For Input Features, select Manhattan Streets. For Clip Features, select UpperWestSide. For Output Feature Class, select UpperWestSideStreets. Run the tool.

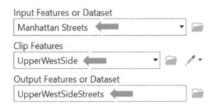

4. **Turn off the original ManhattanStreets layer and save your project.** The result is streets cut cleanly to the Upper West Side neighborhood.

Tutorial 6-3: Merge water features

Sometimes, you must merge two or more layers into a new single layer. For example, you will build one water feature class from several adjacent water feature classes. New York City is made up of five boroughs, each of which is also a county whose water features are downloaded separately from the US Census Bureau.

Open the Tutorial 6-3 project

1. **Open Tutorial6-3.aprx from the Chapter6\Tutorials folder, and save the project as** Tutorial6-3YourName.aprx. The map, NYC Water, includes separate water features for each borough.

2.   **Use the NYC Boroughs bookmark.**

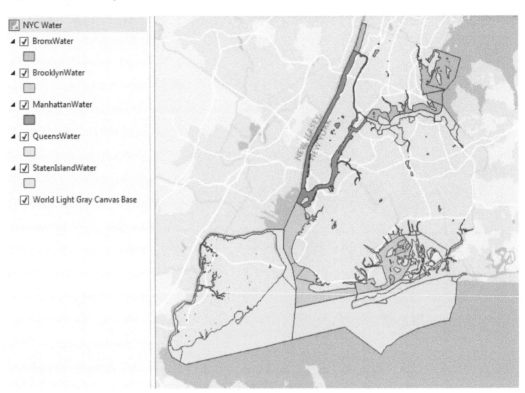

Merge features

Next, you will use the Merge Geoprocessing tool to create one water feature class from five separate feature classes.

1.   **From the Analysis Gallery, open the Merge tool.** The input datasets are the five water datasets as shown, and the Merge Rule is set to First. The output dataset is NYCWater saved to Chapter6. gdb.

2. **Add all five water layers as input datasets: BronxWater, BrooklynWater, ManhattanWater, QueensWater, and StatenIslandWater. For Units, select Feet. Run the tool.**

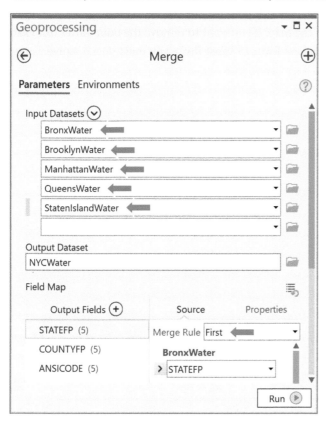

3. **Change the fill color of NYCWater to blue, and zoom to the layer.**

4. **Open the attribute table for NYCWater.** The resulting water feature class has 330 records, one for each water polygon. Merge does not erase boundaries—the polygon boundaries between water features (such as the rivers in this data) are still present. The chosen symbology doesn't draw the boundaries, so they don't appear on the map. If you want to remove the boundaries (for example, to create a single polygon for the East River, or "East Riv"), you must do a dissolve using the FULLNAME field.

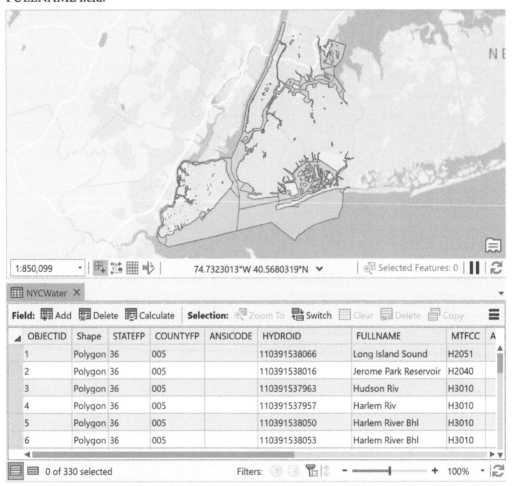

YOUR TURN

Open the NYC Waterfront Parks map, and merge the separate feature classes for each borough into one feature class named **NYCWaterfrontParks**, with fill color green, to Chapter6.gdb. Save your project.

Tutorial 6-4: Append firehouses and police stations to EMS facilities

City agencies may track facility locations in separate feature classes and tables, but an emergency services official wants them in one feature class for better planning. You will use the Append tool to add features to an existing feature class, considering that both have the same attributes (or same schema). In this tutorial, you will add firehouse and police station points to a feature class named EMS.

Open the Tutorial 6-4 project

1. **Open Tutorial6-4.aprx from the Chapter6\Tutorials folder, and save the project as** Tutorial6-4YourName.aprx. The map contains emergency medical service (EMS) facilities.

2. **Zoom to the EMS Facilities layer.**

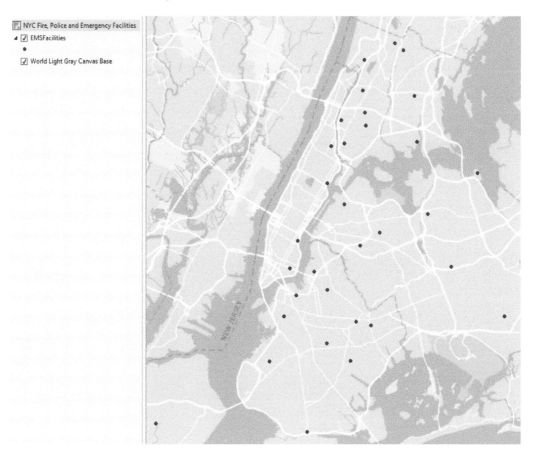

Append features

Here, you will use the Append tool to append firehouses and police stations to already existing EMS points (EMSFacilities). Each layer has the same table (field) structure, also known as a schema. This allows you to choose the option that the input table's schema matches the target table's schema.

1. **Open the EMSFacilities attribute table.** You will see 34 emergency medical facilities.

2. **In the Tools Gallery, open the Append tool.**

3. **For Input Datasets, browse to Chapter 6.gdb, and add FireHouses and PoliceStations.**

4. **For Target Dataset, click EMS Facilities and run the tool.**

5. **In the EMS Facilities attribute table, scroll to see the added firehouses and police stations.**

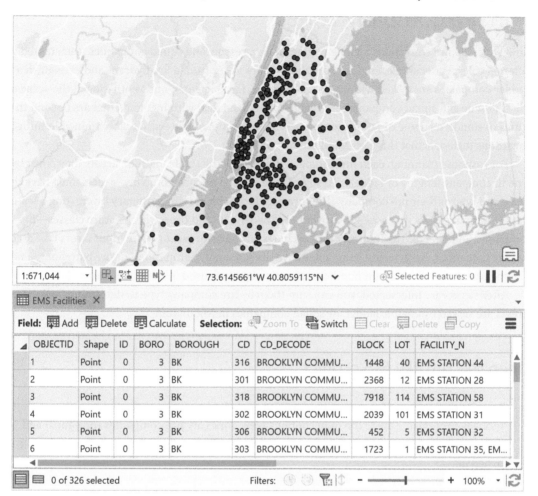

6. **Save your project.**

Tutorial 6-5: Intersect features to determine streets in fire company zones

New York City fire companies include engine, ladder, rescue, and squad companies, each tasked with different roles. For example, engine companies secure water from a fire hydrant and extinguish a fire. Ladder companies are tasked with search and rescue, forcible entry, and ventilation at the scene of a fire. Rescue and squad companies are highly trained and deal with incidents that are beyond the duties of standard engine or ladder companies. Squad companies are also highly trained in mitigating hazardous materials that threaten Manhattan.

For response planning, each fire company must know the total length of streets they cover by the type of company (engine or ladder). To summarize streets for each company, streets must have the name and type of their fire company. The Intersect tool achieves this summary by creating a feature class combining all the features and attributes of two input (and overlaying) feature classes—fire companies and streets. The Intersect tool excludes any parts of two or more input layers that do not overlay each other. Because fire companies have the same number in different fire battalions, a field that includes both the battalion number and fire company is used in the calculations.

After streets are intersected, you can sum them by fire company type to determine how many streets are served by engine, ladder, or squad companies in Manhattan.

Open the Tutorial 6-5 project

1. **Open Tutorial6-5.aprx from the Chapter6\Tutorials folder, and save the project as** Tutorial6-5YourName.aprx. The map contains New York City fire companies (polygons) and Manhattan streets (lines). Fire companies are classified by company type. ManhattanStreets is turned off.

2. **Zoom to the ManhattanFireCompanies layer.**

Open tables to study attributes before intersecting

Studying the attribute tables of each feature class will help you become familiar with the attributes before you intersect features.

1. **Open the ManhattanStreets attribute table, and scroll to the right.** You will find no data about fire companies in the ManhattanStreets feature class, but you will find information about street characteristics, such as the shape (polyline) and length of each street segment.

2. **Open the ManhattanFireCompanies attribute table, and sort ascending by FireCoNum (fire company number).** Scroll to examine the attributes and data of this table. Fields of interest are the shape (polygon) and FireBN_Co_Type, a field that combines fire battalion, company number, and fire company type fields—for example, ladder company (L), engine company (E), or fire squad (Q) are fire company types.

3.   **Close the attribute tables.**

Intersect features

1.   **In the Tools Gallery, open the Intersect tool.**

2.   **For Input Features, select ManhattanStreets and ManhattanFireCompanies. For Output Feature Class, select ManhattanFireCompanyStreets. Save the output features as** ManhattanFireCompanyStreets **to Chapter6.gdb and run the tool.**

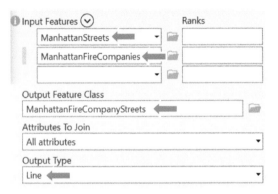

3.   **Turn off all layers except ManhattanFireCompanyStreets and the basemap, use the Central Park bookmark, and zoom in a few times.** The result is street centerlines that include data about the fire companies serving each street.

4. **Click the Explore button, click any of the line features of ManhattanFireCompanyStreets, and scroll to the bottom of the pop-up window.** The result is street centerlines that include data about the fire companies serving each street.

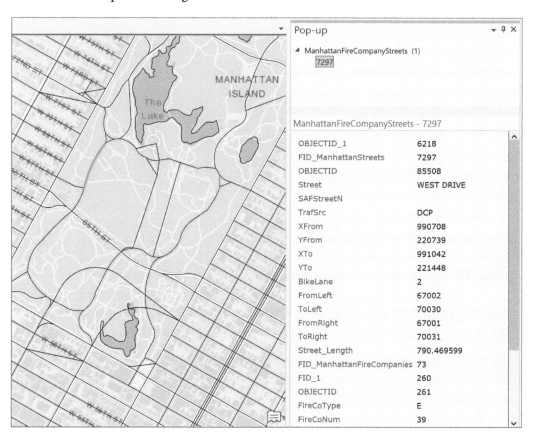

Summarize street length for fire companies

Now that each street segment has fire battalion, company, or type information, you can summarize this information so that emergency planners and fire officials know the total length of streets that each company and fire type serves.

1. **Open the ManhattanFireCompanyStreets attribute table, and right-click FireBN_Co_Type.**

2. **Click Summarize.**

3. **For Input Table, select ManhattanFireCompanyStreets. For Output Table, select FireCompanyTypeStreetLength. For Field, select Street_Length, and for Statistic Type, select Sum. Run the tool.**

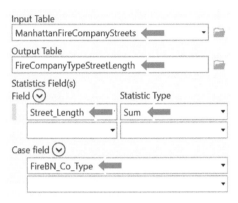

4. **Save the output table as** FireCompanyTypeStreetLength.

5. **Close the ManhattanFireCompanyStreets table.**

6. **Open the FireCompanyTypeStreetLength table.** In the field, SUM_Street_Length, you will see the sum (total length) of street segments for each fire company and type.

OBJECTID_1 *	FireBN_Co_Type	FREQUENCY	SUM_Street_Length
1	1_1_L	139	35732.046198
2	1_10_E	544	96794.841376
3	1_10_L	315	74371.810622
4	1_15_L	532	91798.225658
5	1_4_E	188	35341.360738
6	1_6_E	835	120797.725068

YOUR TURN

Join the FireCompanyTypeStreetLength table to the ManhattanFireCompanies attribute table, open, and sort descending on the field SUM_Street_Length. Select the record (fire company type) with the longest street. Select the next few records to see the location of these fire companies. Close the table, clear all selections, and save your project.

Tutorial 6-6: Union neighborhoods and land-use features

The Union tool overlays the geometry and attributes of two input polygon layers to generate a new output polygon layer. In this tutorial, you will use the Union tool to combine New York City's Brooklyn borough neighborhoods and land-use polygon to calculate the total land area of each type by neighborhood. The output of a union is a new feature layer of smaller polygons, each with combined boundaries and attributes of both neighborhoods and land-use polygons. You can then use Add Geometry Attributes to calculate land-use type (for example, residential) in each neighborhood. Such information would be useful for an urban planner or sustainability director who is interested in learning about house and land-use development in neighborhoods.

Open the Tutorial 6-6 project

1. **Open Tutorial6-6.aprx from the Chapter6\Tutorials folder, and save the project as** Tutorial6-6YourName.aprx. **The map contains Brooklyn neighborhood polygons and land use with zoning types.

2. **Use the Brooklyn Zoomed bookmark.** The features clearly do not share the same boundary, and you will see neighborhoods with mixed zoning types.

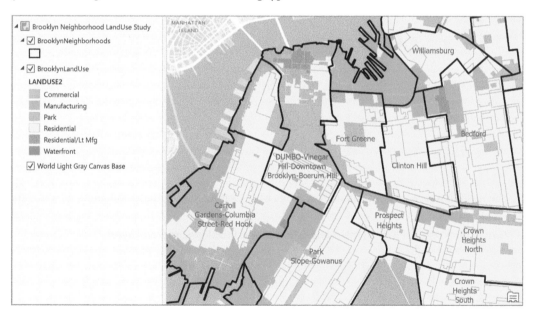

Open tables to study attributes

Studying the attribute tables of each feature class will help you become familiar with the attributes before you use the Union tool. There are 51 neighborhood polygons and 1,294 land-use polygons.

1. **Open the BrooklynNeighborhoods attribute table.** The table contains no data about land use in the feature class, but it does contain data about population and housing units in each neighborhood.

2. **Open the BrooklynLandUse attribute table.** Fields of interest are the land use for each polygon. Both tables include the length and area for every polygon.

3. **Close the attribute tables.**

Union features

1. In the Tools Gallery, open the Union tool.

2. **For Input Features, select BrooklynNeighborhoods and BrooklynLandUse. For Output Feature Class, select BrooklynNeighborhoodsLandUse Union. Run the tool.**

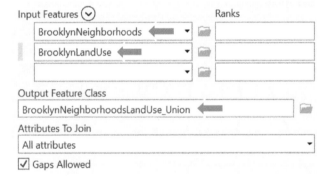

The result is a new layer of smaller polygons with combined neighborhood and land-use data.

3. **Turn off the original BrooklynNeighborhoods and BrooklynLandUse layers.**

4. Symbolize BrooklynNeighborhoodsLandUse_Union using a black outline, width 2, and No Color fill.

5. Zoom in a few times, click one of the new polygons, and see the values for neighborhoods and land use.

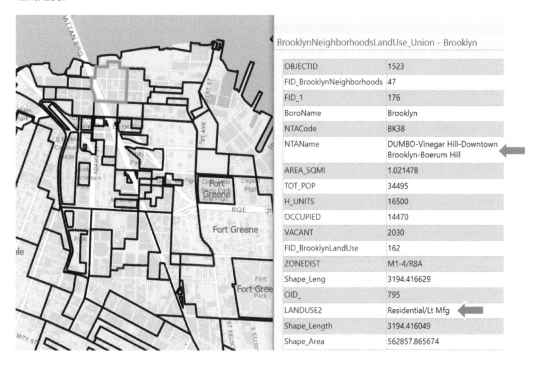

Calculate acreage

A field for the area (acres) of each polygon created by the union is needed before you can summarize the land use for neighborhoods.

1. **Search for and open the** Add Geometry Attributes **tool.**

2.   **For Input Features, select BrooklynNeighborhoodsLandUse_Union. For Geometry Properties, select Area. For Length Unit, select Feet: United States. For Area Unit, select Acres. For Coordinate System, select NAD 1983 StatePlane New York Long Island. Run the tool.**

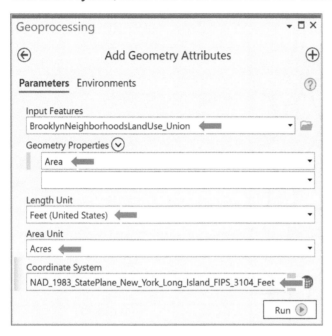

3.   **Open the attribute table for BrooklynNeighborhoodsLandUse_Union, and scroll to the right.**

4.   **Sort descending by POLY_AREA.** The top few polygons are parks followed by residential areas.

Select and summarize residential land-use areas for neighborhoods

1.   **In the Contents pane, right-click BrooklynNeighborhoodsLandUse_Union, and click Properties.**

2.   **Click Definition Query, and create the query:** Landuse2 Is Equal To Residential**.**

3.   **In the attribute table, right-click NTAName, and click Summarize.**

4.   **For Input Table, select BrooklynNeighborhoodsLandUse_Union. For Output Table, select BrooklynNeighborhoodsResidentLandUse. For Field, select POLY_AREA, and for Statistic Type, select Sum. Run the tool.**

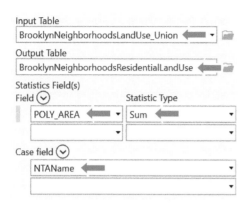

5. **Open the BrooklynNeighborhoodsResidentialLandUse table, and sort descending by SUM_POLY_AREA.**

OBJECTID *	NTAName	FREQUENCY	SUM_POLY_AREA ⌄
21	East New York	39	1490.742844
12	Canarsie	39	1399.285512
2	Bay Ridge	74	1283.27619
44	Sheepshead Bay-Gerrits	15	1283.09128
28	Georgetown-Marine Pai	13	1186.712448
26	Flatlands	16	1127.070164

The list shows neighborhood names with the highest to lowest residential land use.

6. **Using the NTAName field, join the BrooklynNeighborhoodsResidentialLandUse table to BrooklynNeighborhoods.**

7. **Open the table, and sort descending by H_UNITS.** A planner can now compare the neighborhoods with the highest number of housing units and the number of residential acres.

AREA_SQMI	TOT_POP	H_UNITS ⌄	OCCUPIED	VACANT	Shape_Length	Shape_Area	OBJECTID	NTAName	FREQUENCY	SUM_POLY_AREA
1.851587	103169	44583	40676	3907	35635.543552	51619074.711378	15	Crown Heights North	43	1012.613216
2.407514	79371	39069	34951	4118	44407.939371	67117360.743151	2	Bay Ridge	74	1283.275055
1.623299	105804	38816	37190	1626	38737.853254	45254807.739752	25	Flatbush	51	995.75915
1.67394	88727	33939	30774	3165	37137.160582	46666593.573495	5	Bensonhurst West	12	972.76416
4.199814	91958	33772	30806	2966	89196.470873	117083634.748803	21	East New York	39	1490.741526
1.524816	67649	32770	30493	2277	31475.740313	42509247.432791	39	Park Slope-Gowanus	29	706.101339
1.937171	106357	31409	28645	2764	39247.228028	54005019.070944	6	Borough Park	16	1086.358487
2.272107	64518	29706	26236	3470	91047.667084	63342444.994628	44	Sheepshead Bay-Gerr...	15	1283.090145
2.942493	83693	29281	27977	1304	43693.775924	82031668.009968	12	Canarsie	39	1399.284275

8. **Close the attribute tables, and save your project.**

> ## YOUR TURN
>
> Open the Queens Neighborhood Land Use Study map, and repeat the processes to summarize residential land use in the borough of Queens. Save your project.

Tutorial 6-7: Use the Tabulate Intersection tool

Previous tutorials in this chapter used the Intersect and Union tools to create feature classes with combined features and data. With these tools, the data (for example, number of housing units or population) is not apportioned (split into parts and allocated) to the new features. For example, if a single neighborhood crosses more than one land-use zone or more than one fire company, the neighborhood housing data should be split between polygons.

In this tutorial, you use the Tabulate Intersection tool to estimate the number of disabled persons in fire company boundaries using census tracts and fire company polygons. By default, this tool makes apportionments proportional to the areas of split parts of polygons, such as block groups, this method assumes the populations of interest are uniformly distributed by area within polygons.

Open the Tutorial 6-7 project

1. **Open Tutorial6-7.aprx from the Chapter6\Tutorials folder, and save the project as** Tutorial6-7YourName.aprx. The map contains Manhattan census tracts classified with the number of disabled persons (all disabilities) and Manhattan fire companies.

2. **Zoom in, and turn the ManhattanFireCompanies layer on and off to see tracts compared with the borders of fire companies.** Tracts and fire companies clearly do not share exact borders as seen in the graphic.

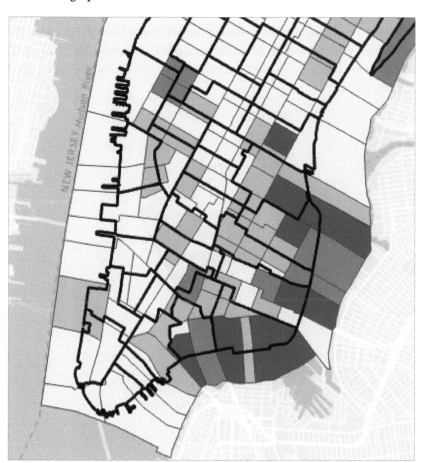

Study tracts and fire company polygons

This exercise uses a small study area composed of four fire companies and 20 census tracts in the Upper West Side to better see the workings of the Tabulate Intersection tool.

1. **Close the Disabled Person Fire Company Study (Manhattan) map, and open the Disabled Person Fire Company Study (Upper West Side) map.** Map labels include Fire Companies 22, 25 (ladder), 74, and 76 (engine). Tracts are labeled with tract IDs (white halo) and the number of disabled persons per tract (yellow halo).

2. **Zoom to Fire Company 76.** Five tracts (selected in the figure) intersect this fire company polygon. Two tracts (019300 and 018900) are completely within the polygon, and their populations are entirely in the fire company. Two tracts (019500 and 019100) are within the polygon but extend into the river, and one (018700) is split between Fire Companies 76 and 22. That tract's disabled person population (880) should be split approximately 50/50, with 440 persons residing in Fire Company 76 and 440 persons residing in Fire Company 22.

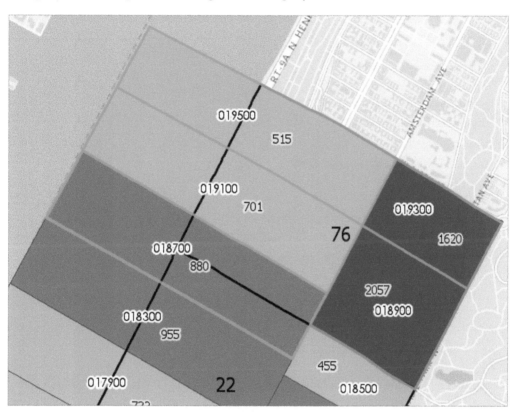

YOUR TURN

Study additional tracts in the other fire company polygons, and identify tracts that are split between the fire company polygons.

Use Tabulate Intersection to apportion the population of disabled persons to fire companies

1. **Search for and open the** Tabulate Intersection **tool.**

2. **For Input Zone Features, select UpperWestSideFireCompanies. For Zone Fields, select FireCoNum. For Input Class Features, select UpperWestSideTracts. For Output Table, select UpperWestSideFireCompanies. For Class Fields, select Tract_ID. For Sum Fields, select Disability. Run the tool.**

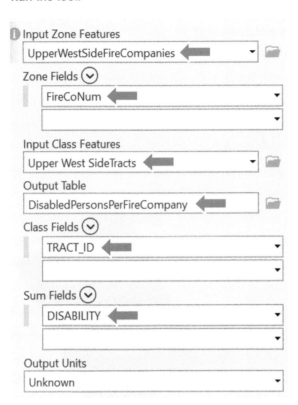

Disregard any warning about tolerance settings.

3. **Open the DisabledPersonsPerFireCompany table, and sort ascending by TRACT_ID.**

4. **Scroll to and select tract 018700.** The population of disabled persons in the tract is indeed split 50/50 between Fire Companies 22 and 76, each with 236 persons. Note that Census tract 019300 (completely within Fire Company 76) has a total population of 1,620 disabled persons.

OBJECTID *	FireCoNum	TRACT_ID ▲	DISABILITY	AREA	PERCENTAGE
8	22	018700	236	0.000012	9.974606 ⬅
24	76	018700	236	0.000012	10.919507 ⬅
9	22	018900	2	0	0.019706
25	76	018900	2055	0.000028	26.845123
26	76	019100	389	0.000024	22.451319
27	76	019300	1620	0.000019	17.789666 ⬅

YOUR TURN

Use Summary Statistics to calculate the total number of disabled persons in each fire company. For Input Table, select DisabledPersonsPerFireCompany. For Output Table, select TotalDisabledPersonsPerFireCompany. For Field, select Disability, and for Statistic Type, select Sum. For Core Field, select FireCoNum. Save your project.

Assignments

This chapter has three assignments to complete that you can download from ArcGIS Online, at go.esri.com/GISTforPro2.8Data:

- **Assignment 6-1:** Build a study area for a rapidly growing Texas metropolitan area.
- **Assignment 6-2:** Use geoprocessing tools to study neighborhoods.
- **Assignment 6-3:** Dissolve property parcels to create a zoning map.

CHAPTER 7

Digitizing

LEARNING GOALS

- Edit, create, and delete polygon features.
- Extract features from other map layers using the study area.
- Create and digitize point and line features.
- Use Cartography tools to smooth features.
- Work with CAD drawings.
- Spatially adjust features.

Introduction

This chapter teaches you how to use heads-up digitizing and basemaps on the computer screen to create vector (point, line, and polygon) map layers. First, you will create a feature class including table attributes. Then you will use an existing map layer, such as Streets, as a guide or spatial data reference for locations. Finally, you will use digitizing tools to create features and enter corresponding attribute data. Technologies such as lidar and satellite imagery are also references for heads-up digitizing. You will learn more about lidar data in chapter 11.

You can use GPS receivers, which collect longitude and latitude data, to create vector features. You can also import computer-aided design (CAD) and building information modeling (BIM) into GIS maps to create feature classes.

In this chapter, you will edit existing features and create features for a rapidly expanding university campus, Carnegie Mellon University (CMU), in Pittsburgh, Pennsylvania. New buildings, additions, and renovations to existing buildings are part of a campus master plan. Changes also include new or modified streets, sidewalks, parking lots, and so on. Architects, engineers, and planners need updated GIS features for the campus, and tutorials in this chapter teach the skills to create and edit them.

Tutorial 7-1: Edit polygon features

In this tutorial, you will move and rotate existing buildings in a map layer. Then you will add vertex points and split polygons to further edit them to match buildings in a World Imagery basemap. Note that imagery maps may change over time so the images may be slightly different from those in the tutorial. The user interface also may look slightly different from the figures you see in this chapter.

Open the Tutorial 7-1 project

1.   **Open Tutorial7-1.aprx from the Chapter7\Tutorials folder, and save the project as** Tutorial7-1YourName.aprx. Tutorial7-1 contains a map of CMU's main campus with buildings as semitransparent features on the World Imagery basemap. Because you will make permanent edits to the features, a clean copy of CMU's building polygon features named Bldgs_Original is located in Chapter7.gdb in case you want to repeat any steps.

2.   **Use the Main Campus bookmark to zoom to CMU's main campus and academic buildings.**

Move features

A few buildings on the campus are not in their correct locations. You can move them into exact locations using World Imagery as a reference.

1.   **Use the Cohon University Center bookmark and Select button to select the Cohon University Center building polygon.**

2.   **On the Edit tab, click the Move button.**

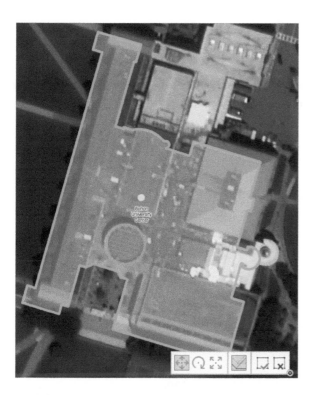

By default, a construction toolbar appears on the active map or when you right-click the map. A yellow dot appears in the center of the polygon indicating that the polygon is ready to edit.

3. **Drag the building polygon to match the building outline in the image.**

4.   **Click anywhere outside the polygon. Continue to drag until the polygon is in the correct location.**

5.   **On the Edit tab, click the Save button, and in the Save Edits pop-up window, click Yes to save your edits.** Saving edits to features differs from saving ArcGIS Pro projects. If you do not want to save changes, click the Discard button.

YOUR TURN

Use the College of Fine Arts bookmark, and move the building polygon to its correct location. You will later add and move vertex points to create the U-shaped building. Pan the main campus map to find other buildings to move to the correct locations. Save your edits and project.

Rotate features

Features may be in the correct location but must be rotated. This requirement applies to CMU's Gates Center for Computer Science.

1.   **Use the Gates bookmark, and select the Gates building polygon.**

2.   **On the Edit tab, click the Move button.** Selecting the Move button enables edits for move, rotate, and scale.

3.   **In the Modify Features pane, click the Rotate button** ⟳. A yellow dot and green circle appear in the center of the building, indicating that the polygon is ready for editing.

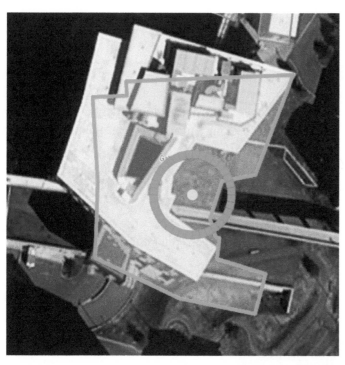

4. **Move the pointer to the green circle, click it, and move and rotate the highlighted outline to match the approximate location of the building outline in the satellite image.** The building may not align perfectly with the satellite image. You will later learn how to edit vertex points to fix an imperfect alignment.

5.   Click anywhere outside the polygon, click the Save button, and click Yes to Save Edits.

Add and move vertex points

Features are sometimes represented as simplified squares, rectangles, or circles, but you can add and modify vertex points to reflect the true building shape. The College of Fine Arts building was drawn as a rectangle, but the building has a U-shaped roof. Adding and editing vertex points will fix this problem.

1.   Use the College of Fine Arts bookmark, and pan the map to see the entire building.

2.   Select the College of Fine Arts building, click the Edit tab, and click the Edit Vertices button ▤. Four vertex points appear at the corners of the polygon. You must add four more vertices to make the U shape.

3.   On the Construction toolbar, click the Add button ▷⁺, and click four points on the left (western) line of the College of Fine Arts building to add four vertex points as shown.

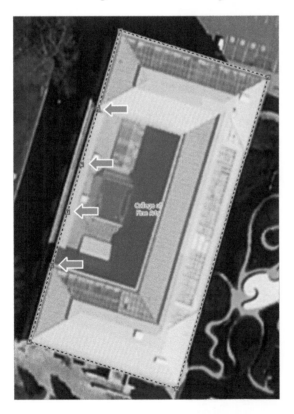

4.   Click the Lasso button and the second of the new vertex points.

5.   Drag the vertex point to the inside corner of the building as shown in the figure.

6. **Click and drag the third vertex point, making a U shape.**

7. **Drag the vertex points to the corner of the building, making a U shape.**

8. **Click the Finish button** .

9. **Save your edits.**

Split features

Two buildings on the campus are drawn as one polygon because they are connected. However, university architects and facility planners must identify the buildings as separate polygons, each with a separate record in the attribute table. The Split tool will accomplish this task.

1. **Use the Baker Porter bookmark, and select the building polygon for these buildings.** Baker Hall is the easternmost section of the building, and Porter Hall connects at the western end of the third wing as shown on the left in the image. This location where the buildings connect is where you will split the polygon.

2. **On the Edit tab, click the Split button** ⊕.

3. **Zoom in, and click the north side of the small indented areas between Baker and Porter Halls, and double-click the south side as shown in the figure. Click and double-click outside the border of the building.** This action splits the building into two polygons at this location.

4. **Save your edits.**

5. **Select the new polygon for Baker Hall, and in the Contents pane, open the Bldgs attribute table.**

6. **Click the Show Selected Records button.**

7. **In the attribute table, type** 2A **for Number,** BH **for BL_ID, and** Baker Hall **for Name.**

8. **Select the Porter Hall polygon, and for its fields, type** 2B **for Number,** PH **for BL_ID, and** Porter Hall **for Name.**

9. **In the attribute table, click Clear to clear the selected records.**

10. **Save your edits, close the attribute table, and use the Baker Porter bookmark.** Two separate buildings are now labeled for Baker Hall and Porter Hall.

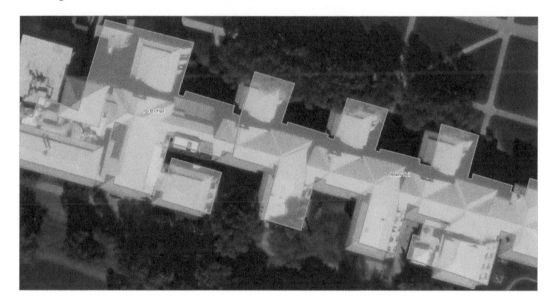

YOUR TURN

Use the Hamerschlag Roberts bookmark to select and split one polygon into two separate build-ings. The rectangular building with two triangles on the left (west) is Roberts Hall, and the building on the right is Hamerschlag. Edit the Roberts Hall polygon using **23** as the building number, **REH** as the BL_ID, and **Roberts Engineering Hall** as the name. Use the Move and Vertices tools to bet-ter match the building polygons to the World Imagery basemap. Save your edits and your project.

Tutorial 7-2: Create and delete polygon features

Changes to the CMU campus include a new addition to the Cohon University Center, a new build-ing for the College of Engineering (Scott Hall), a new quadrangle for the Tepper School of Business (Tepper Quad), and a new addition to Heinz College, which is Hamburg Hall. A new feature class will highlight these development areas for communication with the campus community and the universi-ty's neighbors.

Open the Tutorial 7-2 project

1. **Open Tutorial7-2.aprx from the Chapter7\Tutorials folder, and save the project as** Tutorial7-2YourName.aprx. The tutorial contains a map of CMU's main campus and existing buildings.

2. **Use the Cohon University Center bookmark, and zoom out twice.**

Create a polygon feature class, and add a field

In ArcGIS Pro, you can create feature classes directly in the Catalog pane and add attributes using the attribute table.

1. **Open the Catalog pane, and browse to Chapter7.gdb.**

2. **Right-click Chapter7.gdb, and click New > Feature Class.** The Create Feature Class tool opens, and you have six pages of data to enter.

3. **For Name, type** BldgsProposed; **for Feature Class Type, click Polygon; and click Next.**

4. **Click the Click Here To Add A Field button, type** BLDGNAME, **select Text for Data Type, type** 75 **for Length, and click Next.**

5. **Click NAD 1983 StatePlane Pennsylvania South FIPS 3702 (US Feet).**

6. **Click Next three times, and click Finish.** The new BldgsProposed feature class has been created in the file geodatabase. The feature class is empty now. Later, as you digitize features, polygons will be added to the feature class and become visible on the map.

7. **If the new BldgsProposed layer does not automatically appear in your Contents, add it from the Chapter7.gdb.** Now your new polygon feature class is ready for digitizing.

Add a feature class, and create polygons

Before you start adding features, you'll change the style for the layer's symbol.

1. **In the Contents pane, change the color of the BldgsProposed layer to Tuscan Red with a white outline.**

2. **On the Edit tab, in the Features group, click the Create button** ☑.

3. In the Create Features pane, click BldgsProposed and the Polygon button.

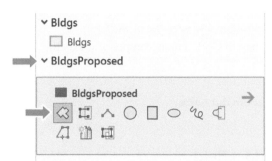

4. On the Configure toolbar, click the Line button ◱, and drag to draw a rectangle that mimics the figure. Double-click the last vertex point to finish the polygon. If the imagery map has an updated building footprint, use that as a guide to digitize the building.

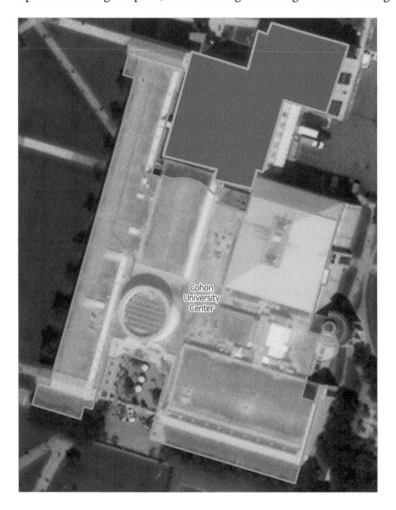

5. Save your edits, and clear the selected polygon.

YOUR TURN

Use the Hamerschlag Roberts bookmark, zoom out a few times, and draw a rectangle on the north of Roberts Engineering Hall. Use the Tepper Quad bookmark to create a polygon using the existing parking lot or building footprint as a guide. Save your edits, and use the Main Campus bookmark to see the proposed developments. In the attribute table, enter building names **Cohon UC Addition**, **Scott Hall**, and **Tepper Quad** to coincide with the names shown in the figure. Show labels using the building name and a white halo. Save your project.

Delete polygons

New development often includes the demolition of existing buildings. As part of the university's Forbes Avenue Innovation Corridor, four buildings will be demolished and must be deleted from the GIS buildings layer.

1. **Use the Arts Park bookmark to zoom to this area of Forbes Avenue on CMU's campus.**

2. **Select one of the two buildings on the right (east) of the Integrated Innovation Institute, and on the Edit tab, click the Delete button ✕, and delete the second building .** You can also right-click after selecting a building, and click Delete. You can use the Shift key to select multiple buildings to delete.

3. **Repeat step 2 to select and delete the two buildings on the right of the 4615 Forbes (GATF) building.**

4.   **Save your edits.**

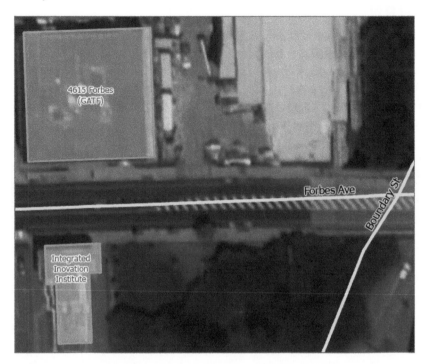

Create a study area polygon feature

Campus architects and campus planners want to know the square footage of a study area of the main campus where most of the academic buildings are located. Planners identify five streets (Forbes, Boundary, Schenley, Tech, and Margaret Morrison) that are used to create the study area. The Trace tool creates a polygon using these streets as guides.

1.   Open the Catalog pane, and expand Databases and Chapter7.gdb.

2.   Right-click Chapter7.gdb, and click New > Feature Class.

3.   Type StudyArea for Name, select Polygon for Feature Class Type, and click Next twice.

4.   Click NAD 1983 StatePlane Pennsylvania South FIPS 3702 (US Feet) as the coordinate system.

5.   Click Next three times, and click Finish.

6.   Change the symbology of the StudyArea layer fill to white.

7.   From the color picker, click Color Properties, and set a 50 percent transparency.

Use the Trace tool to create a polygon feature

1. Use the Main Campus bookmark, and turn off the BldgsProposed and Bldgs layers.

2. On the Edit tab, click the Create button.

3. On the Edit tab, confirm the Snapping button ⊞₊ is on.

4. In the Create Features pane, click the StudyArea layer and the Trace button △.

5. Click the intersection of Boundary St and Forbes Ave. Drag the pointer south along Boundary St, east on Schenley Dr, north and east on Frew St, north and east on Tech St, northeast on Margaret Morrison St, and northwest on Forbes.

6. **Double-click Boundary Street as you near the original point at the intersection of Boundary and Forbes Streets.** The new study area polygon matches the existing street centerlines exactly.

7. Save your edits.

Tutorial 7-3: Create and digitize point and line features

Point and line features can represent public transportation and pathways on CMU's campus. In this tutorial, you will create feature classes for bus stops (points) and directional paths (polylines). You will then digitize points and lines for the bus stops and pathways to two commonly visited buildings on campus, Warner Hall (where the admissions office is located) and the Purnell Center (location of the School of Drama performances).

Open the Tutorial 7-3 project

1. **Open Tutorial7-3.aprx from the Chapter7\Tutorials folder, and save the project as** Tutorial7-3YourName.aprx. The tutorial contains a map of the campus with streets labeled and buildings turned off. Also included in the Contents pane is a crosswalk table of bus stop IDs, stop names, and stop routes.

2. **Use the Bus Stops bookmark.**

Create a feature class and attributes in its table

1. **Open the Catalog pane, and browse to Chapter7.gdb.**

2. **Right-click Chapter7.gdb, and click New > Feature Class.**

3. **Type** BusStops **for Name, select Point for Feature Class Type, and click Next.**

4. **Click the Click Here To Add A New Field button, type** Stop_ID, **select Text for Data Type, type** 5 **for Length, and click Next.**

5. **Click NAD 1983 StatePlane Pennsylvania South FIPS 3702 (US Feet), and click Finish.** A new point feature class is in the file geodatabase and is ready for digitizing. You will use Stop_ID in the next section to join a crosswalk table that contains bus routes associated with each bus stop.

> ### YOUR TURN
>
> Create a Line feature class in Chapter7.gdb named **Paths**. Create a field named **PathName**, with Text for Data Type and **25** for Length, using coordinate system NAD 1983 StatePlane Pennsylvania South FIPS 3702 (US Feet).

Digitize points

1. In the Contents pane, rename the new point feature class **BusStops** CMU Bus Stops. This new name reflects the stops for CMU's main campus area that you will digitize next.

2. **Change the symbol to Bus Station, color to Ultra Blue, and size to** 18.

3. **On the Edit tab, click the Create button.**

4. **In the Create Features pane, click CMU Bus Stops and the Point button.**

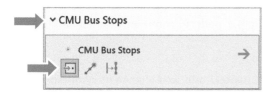

5. **Working west to east, click points on the map to digitize bus stops 01 to 15 at the approximate locations in the order as labeled in the figure.**

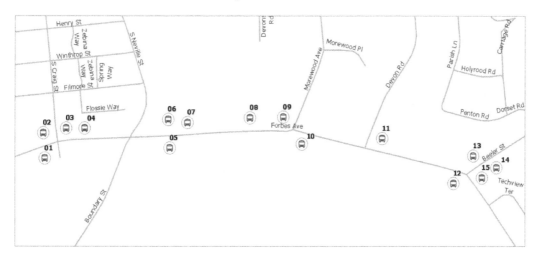

6. **Close the Create Features pane.**

7. **Open the attribute table, and type the Stop_ID values as** 01, 02, 03, **and so on to match the labels on the map.** This field is used to join the crosswalk table with bus routes associated with each stop. You will label the map in Your Turn.

8. **Save your edits.**

YOUR TURN

Join the BusStopCrossWalk table to CMU Bus Stops using the Stop_ID fields. On the map, click bus stop 09 at the intersection of Forbes and Morewood Avenues to see the stop name and route. Close the Create Features pane and attribute tables, and clear selections. Label the map using the Stop_ID field.

Digitize pathways

Street centerline and sidewalk/curb layers don't provide internal campus paths to direct visitors from bus stops or a parking garage. In this exercise, you will digitize paths to commonly visited buildings on CMU's campus.

1. If it isn't already listed in Contents, add Paths from Chapter7.gdb, and change its symbology to Mars Red, size 2 pt.

2. Add the Streets basemap, and use the Paths bookmark.

3. On the Edit tab, click the Create button.

4. In the Create Features pane, click Paths and the Line button.

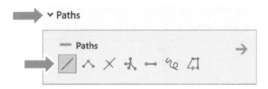

5. Click a starting point at bus stop 10, click points along the path as shown in the image, and double-click to end the path approximately at the entrance of Warner Hall.

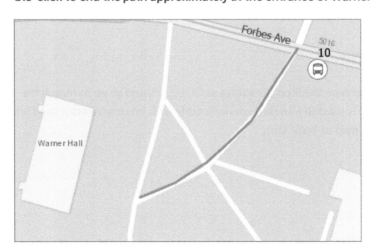

6. **Open the Paths attribute table, and type** Bus to Warner Hall **under PathName.**

7. **Digitize a path from Warner Hall to the Cohon University Center, and type** Warner Hall to CUC
    for the path name in the attribute table.

8. **Save your edits, clear the selection, and label the paths using the PathName field, using a font
    size** 12, **bold font, and Basic Line placement.**

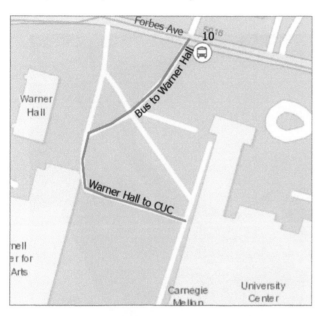

YOUR TURN

Digitize the paths from a bus stop to the Purnell Center building (just south of Warner Hall), and
enter path name **Bus to Purnell Center**. Close the Edit and attribute tables, and clear selections.
Save your project.

Tutorial 7-4: Use Cartography tools

In addition to using the Edit tools, you can modify GIS features using Cartography tools. A useful tool to improve the aesthetic or cartographic quality of polygons is the Smooth Polygon tool.
Organizations such as the US Census Bureau sometimes digitize lines and polygons at a small
scale. The features are digitized with just a few line segments and don't match the true geography.
Smoothing features can fix this problem.

Open the Tutorial 7-4 project

1. **Open Tutorial7-4.aprx from the Chapter7\Tutorials folder, and save the project as** Tutorial7-4YourName.aprx.

2. **Use the Flagstaff Hill & Panther Hollow Lake bookmark.** This tutorial contains a map with semi-transparent polygon features, a park adjacent to the university's campus, a golf course next to campus, and a lake south of campus. The map is zoomed to the park and lake.

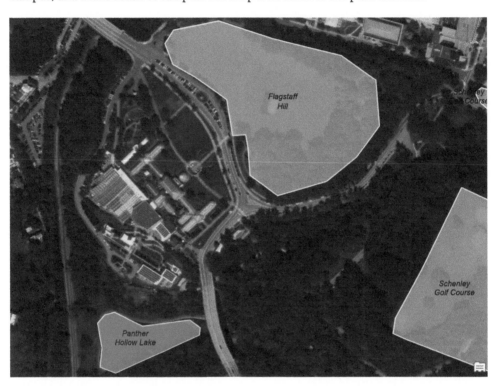

Smooth a green space polygon

Flagstaff Hill is part of Pittsburgh's Schenley Park between Carnegie Mellon University's campus and Phipps Conservatory. The Flagstaff Hill polygon was roughly drawn with just a few line segments and can be smoothed.

1. **On the Analysis tab, click the Tools button.**

2. **In the Geoprocessing pane, click Toolboxes, and expand Cartography Tools > Generalization, and click Smooth Polygon.**

3. **For Input Features, select Greenspaces. For Output Features, type** GrenspacesSmoothed. **For Smoothing Algorithm, select Polynomial Approximation With Exponential Kernel (PAEK). For Smoothing Tolerance, type** 150 **and select Feet. Save the output feature class to Chapter7.gdb.**

The Smoothing Tolerance is important when creating the vertices of the new path. A shorter length will result in a more detailed (or smoother) path but will take longer to process.

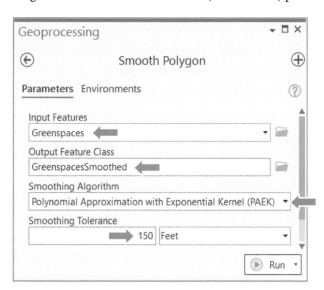

4. **Run the tool.**

5. **Turn off the Greenspaces and Water layers, and zoom to the GreenspacesSmoothed layer.** The result is a new feature class of smoothed polygons for Flagstaff Hill and the polygons for Schenley Golf Course.

6.   **Turn the Water layer on.**

YOUR TURN

Use the Smooth Polygon tool to smooth the water feature of Panther Hollow Lake. Save the new features as **WaterSmoothed** to Chapter7.gdb. Turn off the Water layer, and save your project.

Tutorial 7-5: Transform features

Administrators, architects, and facility planners often combine GIS map layers and CAD drawings of separate buildings for strategic planning, including understanding the occupancy and use of every space on campus. CAD drawings, however, use Cartesian coordinates and are not geographically referenced to GCS, state plane, or UTM coordinates. CAD drawing units are also different from GIS units. For example, the leading CAD software, AutoCAD, uses one inch or one millimeter as the unit, and GIS maps generally use feet or meters as the unit. Transforming features in GIS makes aligning CAD drawings to GIS maps easy, regardless of drawing or map coordinates and units. In this tutorial, you will import one campus building, Hamburg Hall, and will use Move, Rotate, and Transform tools to place the CAD drawing of Hamburg Hall in its correct geographic location on a campus map.

Open the Tutorial 7-5 project

1.   **Open Tutorial7-5.aprx from the Chapter7\Tutorials folder, and save the project as** Tutorial7-5YourName.aprx.

2.   **Use the Study Area Buildings bookmark.** The project contains a map of CMU's campus and the Oakland neighborhood to the north of campus. Academic buildings for the main part of campus are included and used to transform CAD drawings.

Add and export a CAD drawing

Next, you will add an AutoCAD drawing that includes polygons for every space on the first floor of CMU's Hamburg Hall academic building.

1.   **Click Add Data, browse to Chapter7 > Data, double-click HBH1.dwg, click Polygon, and click OK.**

2.   **Zoom to the HBH1-Polygon Group layer.** The drawing unit is one inch, and the drawing appears as 12 times its actual size.

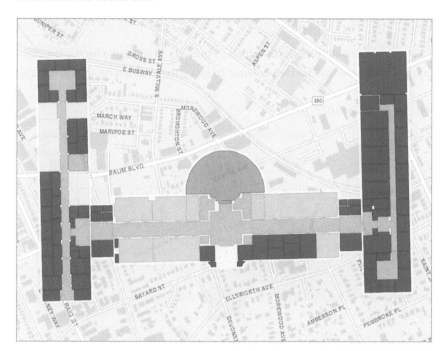

3.   **In the Contents pane, click the arrow next to HBH1-Polygon Group to expand the layers.** CAD drawings contain layers in one drawing and are color-coded according to the layer color assigned in the CAD drawing. You cannot edit CAD drawings directly, so you will export the drawing to a feature class.

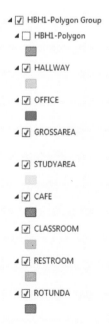

4. **In the Contents pane, right-click layer HBH1-Polygon, and click Data > Export Features.**

5. **In the Geoprocessing pane, for Output Name, type** HBH1SpacePlan. **Save it to Chapter7.gdb.**

6. **Run the tool.** HBH1SpacePlan is automatically added to the map in the same location as the original CAD drawing. The new layer will be transformed to its actual location on the campus map.

7. **Remove the HBH1-Polygon Group.**

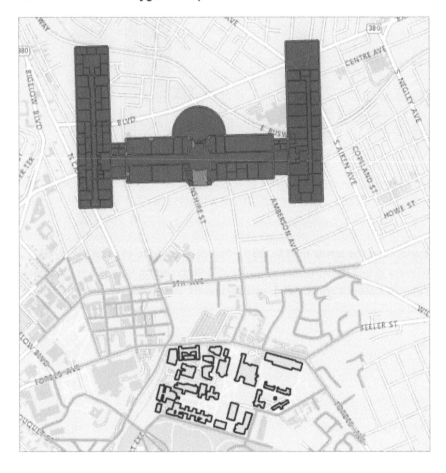

Explore attributes, and classify layers

Before transforming the drawing to the GIS coordinates, you will explore the attributes of the exported CAD drawing and classify spaces based on their layer assignment.

1. **Open the HBH1SpacePlan attribute table.** The results of exporting a CAD drawing are that properties of the drawing are added as fields in the attribute table. The Layer field shows designations of space use. Every polyline in AutoCAD became a polygon in the new feature class, and the layer on which the CAD polyline resided became the layer designation (for example, Office, Hallway, Classroom, and so on).

OBJECTID	Shape	Entity	Layer	LyrFrzn	LyrOn	Color	Linetype	Elevation	LineWt
1	Polygon	LWPolyline	HALLWAY	0	1	40	Continuous	0	25
2	Polygon	LWPolyline	HALLWAY	0	1	40	Continuous	0	25
3	Polygon	LWPolyline	OFFICE	0	1	5	Continuous	0	25
4	Polygon	LWPolyline	OFFICE	0	1	5	Continuous	0	25
5	Polygon	LWPolyline	HALLWAY	0	1	40	Continuous	0	25
6	Polygon	LWPolyline	OFFICE	0	1	5	Continuous	0	25

2. **Close the attribute table, and in the Contents pane, right-click HBH1SpacePlan, and click Symbology.**

3. **From the Options menu on the top right of the Symbology pane, select Import Symbology. You could also search for the tool Apply Symbology From Layer.**

4. **In the Geoprocessing pane, under Symbology Layer, browse to Chapter7\Data\, and click SpacePlan.lyrx.**

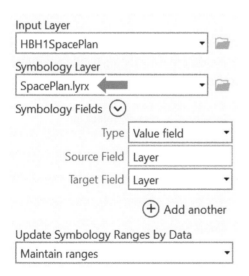

5. **Run the tool.**

6. **In the Contents pane, turn off the World Light Gray Canvas Base and Streets layers.**

7. **Zoom and pan the map to see the HBH1SpacePlan and study area buildings.**

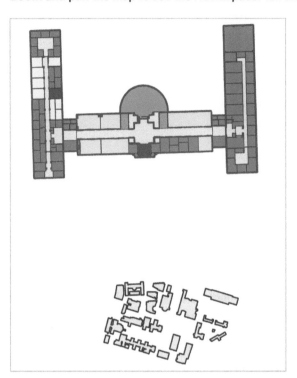

Rotate a building

The features are now ready for transformation into the GIS units for the map. First, you will rotate the Hamburg Hall building 180 degrees. Then you will move it closer to the existing building outlines for campus, near Hamburg Hall. This will make the transformation easier.

1. **In the Contents pane, right-click HBH1SpacePlan > Selection, and click Select All.**

2. **On the Edit tab, expand the Tools Gallery.**

3. **In the Alignment group, click the Rotate button.**

4. **Drag the green circle to rotate the building approximately 180 degrees.**

5. **Pan the map if necessary to see the entire building.**

6. **Move the polygons closer to the study area buildings.**

7. **Save your edits, and zoom in to see your modified building and study area buildings.** Hamburg Hall is the second building from the top left. The new Hamburg Hall layer that includes the space plan layers will be aligned to the existing outline of Hamburg Hall on the campus map.

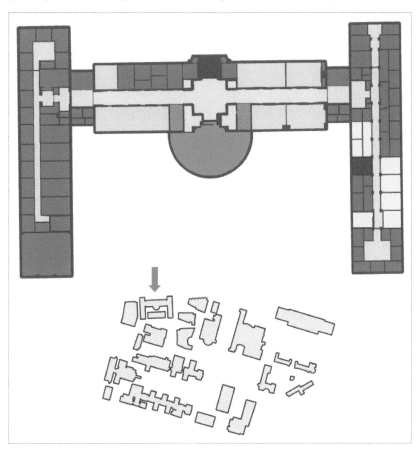

Transform polygons

Next, you will transform the building using eight links that align the floor-plan polygons to the building in the GIS map. Such a transformation will match the Hamburg Hall layer with space plan details to its correct geographic location on the map.

1. In the Contents pane, right-click HBH1SpacePlan > Selection, and click Select All.

2. In the Modify Features pane, click the Transform button.

3.   In the Modify Features pane, under Transformation Method, click Similarity, and click the Add
     New Links button.

4.   Click the lower-left corner of the HBH1SpacePlan layer and the corresponding lower-left cor-
     ner of the Hamburg Hall building in the campus map.

5. **Continue adding links as shown in the figure. Use your wheel button to zoom in and out while adding points, clicking the wheel button to pan.**

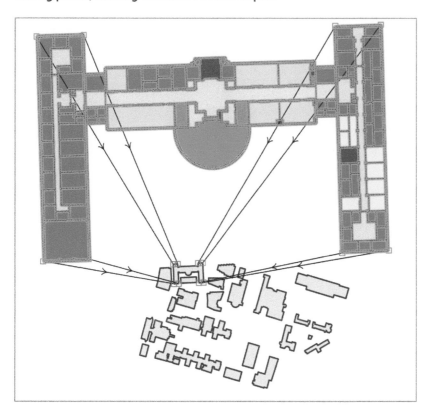

6. **In the Modify Features pane, click the Transform button in the lower-right corner.**

7. **Wait while the polygons are scaled and moved to the geometry on the map.**

8. **Save your edits, and clear selected features.**

9. **Zoom to the HBH1SpacePlan layer to better see the transformed polygons on the map.**

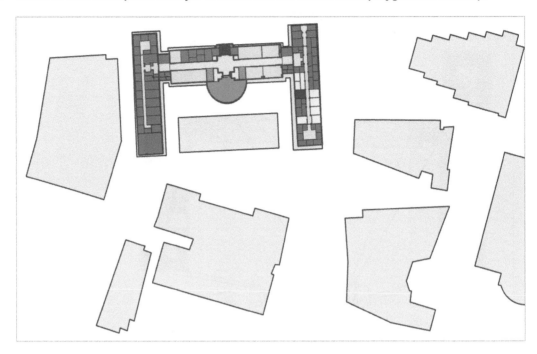

10. **Save your project.**

Assignments

This chapter has two assignments to complete that you can download from ArcGIS Online, at go.esri.com/GISTforPro2.8Data:

- **Assignment 7-1:** Conduct a campus storm water runoff project.
- **Assignment 7-2:** Digitize police beats.

Geocoding

LEARNING GOALS

- Get an overview of the geocoding process.
- Geocode using zip codes.
- Geocode addresses using streets.

Introduction

Geocoding is a GIS process that matches location fields in tabular data to corresponding fields in existing feature classes to map the tabular data. An example is the survey data from this chapter that lists street addresses and zip codes of the people surveyed. Using these location fields, you can geocode their locations using zip code polygon or street feature classes. Notice here that geocode is a verb; for example, use a geoprocessing tool to geocode the data. But when referring to fields for a location field, such as zip code, it's a noun; for example, the ZIP field in the table is a geocode. Another example is transaction data collected by organizations. Because transaction data records, such as for deliveries of appliances, often include location fields, it's possible and useful to map these locations—for example, to optimize routing of delivery trucks that have several stops to make. Another example of transactions is patients in a hospital. In this case, it's useful to map the residences of patients to identify the service area of the hospital.

The source data that you will geocode in the beginning of this chapter is from a survey taken by an arts organization from attendees of its annual art show. The arts organization wants to analyze locations of attendees to better target future marketing efforts. You'll geocode first for a multistate region by zip code and then for the home county of the arts organization by street address.

One problem with geocoding is that source data suppliers (for example, survey respondents for the survey data) and data entry workers can write or type anything they want for an address, including misspellings, abbreviations, omissions, and place-names such as "University of Pittsburgh" instead of an address. Consequently, exact matching of sources to reference data is not possible. Instead, GIS must use fuzzy matching and make matches that are approximate instead of always 100 percent accurate. For example, the address "123 Fleet St" may be on a reference street map, but a surveyed person may have written "123 Fleat" for source data, with a misspelling of "Fleet" and without the "St." A fuzzy-matching algorithm nevertheless may determine that the source address is close enough to the reference street address and plot the residence at 123 Fleet St.

A rule-based expert system is software that makes fuzzy matches, and the geocoding software in ArcGIS Pro is such a system. The system attempts to use the thought processes and rules that an expert would use to accomplish a complex and ambiguous task. For example, you can think of the geocoding expert system as attempting to mimic what a resourceful mail delivery person would do, using their expert knowledge to get a badly addressed piece of mail to the right address. The following expert system components are used in ArcGIS Pro:

- *Source table*, including geocodes (in this chapter, street addresses and zip codes) to be mapped
- *Reference data*, which has mapped features such as street centerlines or zip code polygons and fields such as street names that can be fuzzily matched to the source table's geocodes
- *Geocoding tool,* with the algorithms, rules, and user interface to perform geocoding
- *Locator*, a reusable set of files that include all geocoding parameter values and data for a given kind of reference data

To account for spelling errors, an algorithm computes a Soundex key, which is a code assigned to names that sound alike (for example, "Fleet" and "Fleat" both have Soundex key F43), and identifies candidate matches of source and reference street addresses. The algorithm starts with a score of 100 for each source record and subtracts penalty points for each problem encountered. If the end score is greater than the minimum candidate parameter value set by default or by the user, a reference location is a candidate for matching. If there are two or more candidates, the candidate with the maximum score is chosen as the estimated location. If there is a tie, one of the tying locations can be arbitrarily assigned as the match unless the user chooses to not accept ties.

In this chapter, you will geocode using zip codes and street centerlines. People generally will disclose their zip codes in surveys and get them right, so geocoded results generally are complete and accurate, albeit only at the zip code level. Often, a zip code may be the only available data type and will suffice for marketing purposes. Service or product delivery and other location-based needs require more precise locations. Street centerlines are sufficient for many purposes, but certainly not for locating in-ground natural gas lines during construction digging. You can geocode with zip codes and street centerlines with free map layers downloaded from the internet (see chapter 5). However, cities and states perform many other kinds of geocoding often more precisely, for example, using land parcel centroids with street addresses as provided by many city governments. Note that Esri provides the highly accurate and current ArcGIS Online World Geocoding Service. If you are in a class, however, ask your instructor before using this service, because using the service consumes credits that must be purchased from Esri.

Street centerline maps, available from the Census Bureau's TIGER/Line data and from vendors, are widely used for geocoding street addresses but are limited by only having house numbers on the left and right for the beginning and end of each one-block-long street segment. Consequently, addresses within blocks are linearly interpolated (for example, 150 Main St. is plotted halfway along the street segment with ranges 100 to 198 and 101 to 199) and are not exact locations.

Generally, not all source data records are matched when geocoding. A performance measure for geocoding is the percentage of source addresses that get matched and plotted using the reference data. To compute match rates, subtract all records in source data that do have not addresses (location fields that are blank, do not start with a house number, are not street intersections, and so on) from the total number of addresses in the source data used as the denominator for the match rate.

Unfortunately, there is no way to judge how many matches are truly correct. Organizations that critically depend on geocoding (such as 911 emergency calls for services from police, fire, and ambulance responders) review nonmatches and incorrect matches to improve their maps and procedures for obtaining correct source data from callers.

Tutorial 8-1: Zip code geocoding

In this tutorial, you will geocode survey data collected by a Pittsburgh arts organization that holds an event each year attended by persons across the country, but mainly by those residing in the four-state region of Pennsylvania, Ohio, Maryland, and West Virginia. You can use such survey data, if geocoded, for marketing, philanthropy, or other means of communication with its patrons.

Open the Tutorial 8-1 project

1. **Open Tutorial8-1 from Chapter8\Tutorials, and save the project as** Tutorial8-1YourName.

2. **Use the Region bookmark.** The AttendeesPARegion.csv table is the source data for geocoding, and PARegionZIP (zip code polygons) is the reference data. AttendeesPARegion.csv has a numeric ZIPCode field, and reference layer PARegionZIP has zip code data saved as text in its GEOID10 field. Some zip codes of the region start with 0, which were not included in the source data's numeric version, so the remedy was to calculate ZIPCodeNum from GEOID10 in the reference data, which exists and is a numeric field that drops leading zeros.

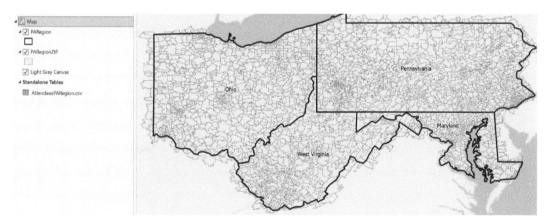

Build a zip code locator

Recall that a geocoding locator is a set of files that stores parameters and other data for the geocoding process.

1. **Search for and open the** Create Locator **tool.**

2. **Complete the tool as follows.** Note that selecting the ZIP field for the locator's Role makes this a
    ZIP Code Locator.
    - **Country or Region:** United States
    - **Primary Table(s):** PARegionZIP; **Role:** ZIP
    - ***ZIP:** ZIPCodeNum
    - **Output Locator:** PARegionZIP_CreateLocator
    - **Language Code:** English

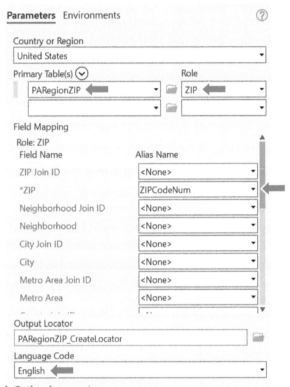

3. **Run the tool.**

4. **Open the Catalog pane, expand Locators, right-click PARegionZIP_CreateLocator, click Locator Properties, and click Geocoding options.** These are the default geocoding parameters used for this zip code locator—namely, minimum match score and minimum candidate score. For each matching problem detected, the geocoding algorithm subtracts penalty points from a starting score of 100. When the algorithm finishes scanning for problems, if the scores are 60 or larger, as set here by default, the minimum candidate and match score rules are met. You can change the minimum match score or minimum candidate score threshold scores to adjust (tune) the fuzzy-matching process and make the process more conservative or liberal. High values are conservative (fewer match errors allowed), and low values are liberal (more match errors). You'll use the defaults for geocoding the arts organization survey data.

5. **Close the Locator Properties window, and hide the Catalog pane.**

Geocode data by zip code

1. **Open the AttendeesPARegion.csv table.** The table has 1,123 survey responses, and if you sort by ZIPCode and scroll down, you'll see no records with missing zip code values. Records with missing zip codes were deleted.

2. **Close the table.**

3. **Search for and open the** Geocode Addresses **tool.**

4. **In the Geoprocessing pane, complete the tool parameters as follows.**
    *   **Input Table:** AttendeesPARegion.csv
    *   **Input Address Locator:** PARegionZIP_CreateLocator
    *   **Input Address Fields:** Multiple Field
    *   **Address or Place:** Address
    *   **City:** City
    *   **State:** State
    *   **ZIP:** ZIPCode
    *   **Output Feature Class:** Attendees
    *   **Output Fields:** All

Important note: Do not select ARCGIS World Geocoding Service for Address Locator unless you have permission from your instructor or employer. The organization account you are using will be billed for using the geocoding service and you may be billed.

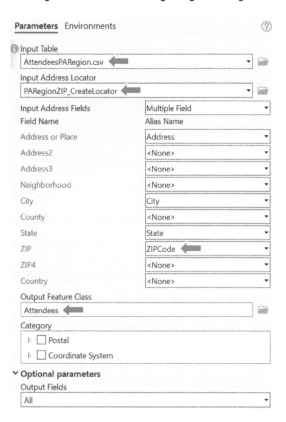

5.   Click Run, but don't close the Geoprocessing pane after it finishes.

6.   At the lower left of the Geoprocessing pane, click View Details > Messages to see that 1120 out of 1123 (99.73%) attendee records were matched. Given that all input records had zip codes and zip code data is generally accurate, this high match rate is as expected.

7.   Close the Geocoding Addresses and Geoprocessing panes, and symbolize Attendees with a red circle 3, size 5 pt. Each matched source address is plotted at its zip code centroid. Although only one point is visible per zip code, generally many points are on top of each other for all attendees from the zip code area.

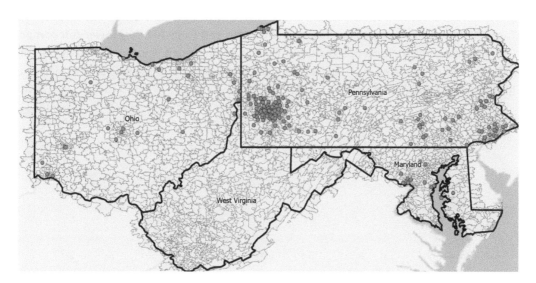

8. In the Contents pane, open the Attendees attribute table and sort Status Descending. The three unmatched records have street addresses, cities, and states, so you can look up zip codes at the US Post Office website (search for zip code lookup in your web browser). The first unmatched record has an incorrect zip code value in the survey, 15230. The value should be 15213. The other two records have correct zip code values, but the reference data, PARegionZip, does not have polygons for those zip codes. The US Census Bureau's zip code maps are approximations constructed from US Postal routes, and likely the two missing zip code polygons are the result of construction methods for estimating zip code polygons.

9. Close the table, and save your project.

Rematch attendee data by zip code

The match rate, 99.7 percent, is extremely high and well above the threshold for any marketing decision-making or other management purposes, so you can use the Attendees map without any changes. But for practice, you'll rematch to 100 percent. You'll correct the zip code in one record and pick approximate points on the map for the two zip codes that are not in the reference data.

1. In the Contents pane, right-click Attendees, and click Data > Rematch Addresses. The first unmatched record comes up with the incorrect zip code, 15230. In the next step, you'll correct the zip code value as part of the rematching process.

2. In the Rematch Addresses pane, for ZIP, type 15213, press TAB. The new zip code yields a candidate with a score of 73, which is above the threshold of 60.

3. Click the Match button ✔. That record is now matched and mapped.

4. Drag the Rematch pane away from the ArcGIS Pro window so that you can see the map and contents.

5. **Turn off PARegionZIP, change the basemap to Streets, and zoom in to the lower-left corner of Pennsylvania and northwest of Pittsburgh.** There is no problem if you cannot find the location indicated on the following map. Any location will do for the sake of learning how to pick a point from a map for geocoding.

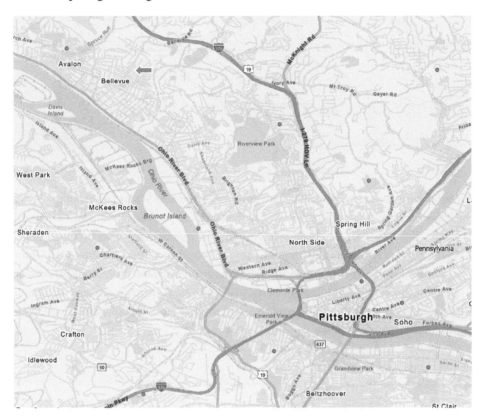

6. **In the Rematch Addresses pane, click the Pick From Map button** **, and click the map approximately at the point directed by the arrow, as shown in the figure in step 5.**

7. **Click the Match button.** The record is matched. If your map zooms in, click the Previous Extent button to see that the picked point was matched and is on the map.

8. **Use the Region bookmark, and click the Pick From Map button to deselect it.**

> ## YOUR TURN
>
> Similarly, match the remaining unmatched record to any location in eastern Pennsylvania. This work is just for practice, so it's not important which point you pick. Save your edits.

Symbolize using the Collect Events tool

Now you have attendees' survey data geocoded to zip code centroids, generally with many attendees at centroids. For symbolizing attendees, next you will count the number of attendees in each zip code and plot graduated symbols, with symbol size increasing as the number of attendees increases. You can do this work manually in a couple of steps, but the Collect Events tool does the job in one step.

1.   **Turn off Attendees, change the basemap back to Light Gray Canvas, and use the Region bookmark.**

2.   **Search for and open the** Collect Events **tool.**

3.   **Select Attendees for Input Incident Features, and leave the output name as the default value.**

4.   **Run the tool.**

5.   **Zoom in to the southwest corner of Pennsylvania where Pittsburgh is located.**

Although you can change the symbology of the output perhaps to better portray the spatial distribution of attendees, Collect Events has done the hard work of counting by zip code and applying graduated symbols.

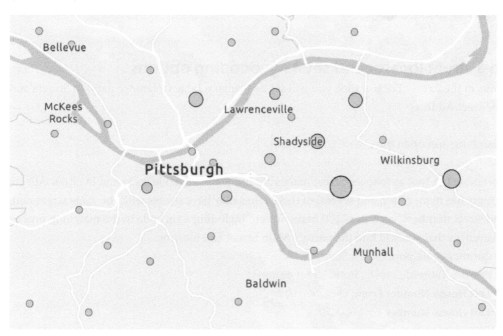

6.   **Save your project.**

Tutorial 8-2: Street address geocoding

This tutorial starts with the data from tutorial 8-1 but only includes records that have street addresses and are in Allegheny County, which includes the city of Pittsburgh, where the arts event is held each year. Allegheny County is the local market for the arts event, and more detailed location data on attendees is desirable for marketing there. So you'll geocode by street address to place unique points on the map for attendees in the county. You'll use the same workflow as with zip code matching: build a locator (but this time using street centerlines as the reference data) and geocode the source data of survey respondents. As a step for tuning the geocoding process, however, you will set very low values for the minimum match and candidate score parameters of the locator, namely 10 and 10, replacing the defaults of 60 and 60. As a result, geocoding will make obvious errors for low parameter values. By analyzing the attribute table of the geocoded output, you can select values for the minimum match and candidate score parameters that are best for the survey data. Scores below the ones you select will make identifiable match errors, and those above will be accurate enough for marketing purposes.

Open the Tutorial 8-2 project

1.　**Open Tutorial8-2, and save the project as** Tutorial8-2YourName.

2.　**Use the Allegheny County bookmark.** You are seeing 92,430 block-long street segments on your map of Allegheny County.

Build a street locator and set its geocoding options

This time, in the Create Locator tool, you will set its Primary Table (reference data) to Streets and its Role to Street Address.

1.　**Search for and open the** Create Locator **tool.**

2.　**Complete the tool as follows.** Street names are unique in zip code areas and in cities. Allegheny County has many cities, and several of these cities may have streets with the same street name and street numbers, such as a "100 Main Street." Including a zip code in the matching process guarantees that you will find the correct Main Street and location.
   - **Country or Region:** United States
   - **Primary Table(s):** Streets; **Role:** Street Address
   - ***Left House Number From:** LFROMADD
   - ***Left House Number To:** LTOADD
   - ***Right House Number From:** RFROMADD
   - ***Right House Number To:** RTOADD
   - **Street Name:** FULLNAME
   - **Left ZIP:** ZIPL
   - **Right Zip:** ZIPR

- **Output Locator:** Streets_CreateLocator
- **Language Code:** English

Parameters Environments   ⑦

Country or Region

| United States | ▾ |

Primary Table(s) ⌄   Role

| Streets ⬅ | ▾ | 📁 | | Street Address ⬅ | ▾ |
| | ▾ | 📁 | | | ▾ |

Field Mapping

*Left House Number From	LFROMADD ⬅	▾
*Left House Number To	LTOADD ⬅	▾
*Right House Number From	RFROMADD ⬅	▾
*Right House Number To	RTOADD ⬅	▾
Left Parity	\<None>	▾
Right Parity	\<None>	▾
Prefix Direction	\<None>	▾
Prefix Type	\<None>	▾
*Street Name	FULLNAME	▾
Suffix Type	\<None>	▾

Output Locator

| Streets_CreateLocator | 📁 |

Language Code

| English ⬅ | ▾ |

› **Optional parameters**

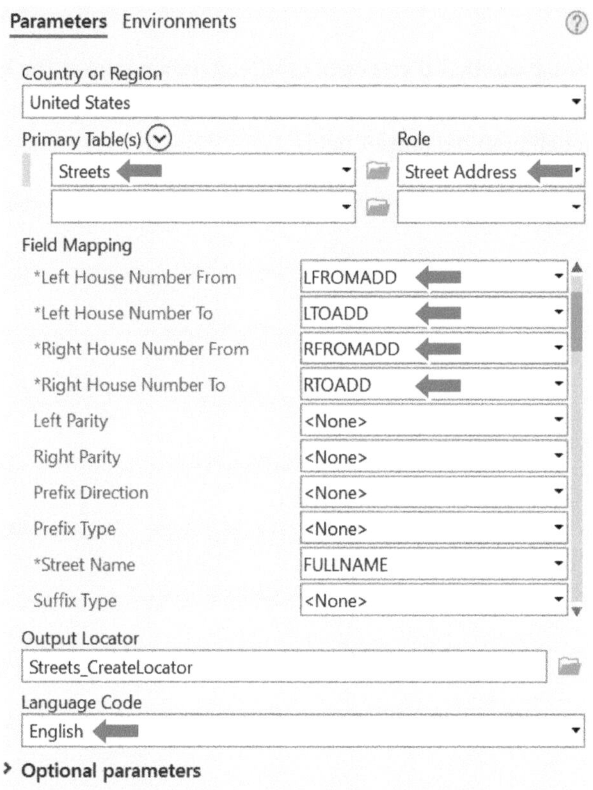

3. **Run the tool.**

4. **Open the Catalog pane, expand Locators, right-click Streets_CreateLocator, and click Locator Properties.**

5. **In the Locator Properties pane, click Geocoding options.**

6. **For Minimum match score and for Minimum candidate score, type** 10 **for each and click OK.**

Geocode attendee data by street address

1. **Search for and open the** Geocode Addresses **tool.**

2. **Complete the tool parameters as follows.** Note that if you click Optional parameters at the bot-
    tom of the tool, you can change the number of output fields in the geocoded output feature class.
    Leave All as the selection so that you get all possible output fields.
    - **Input Table:** AttendeesAlleghenyCounty.csv
    - **Input Address Locator:** Streets_CreateLocator
    - **Input Address Fields:** Multiple Field
    - **Address or Place:** Address
    - **City:** City
    - **State:** State
    - **ZIP:** ZIP_Code
    - **Output Feature Class:** AttendeesAlleghenyCounty_Geo1010

3. **Run the tool and when it finishes, click View Details > Messages.** A total of 912 of 932 records are
    matched, for a match rate of 97.85 percent. Of the remaining records, 10 are unmatched. Also, 10
    are tied and plotted. Note that ArcGIS software developers often tweak the geocoding algorithm
    in new software releases, so your match rate may be slightly different.

4. Close the Geocode Addresses window and the Geoprocessing tool.

Select minimum candidate and matching scores

Here you will rearrange the order of fields displayed in the geocoded results attribute table and then sort the table. Then you can compare the input attendee addresses and zip codes versus those matched by geocoding.

1. Open the attribute table for AttendeesAlleghenyCounty_Geo1010.

2. In the upper-right corner of the attribute table, click the menu button, deselect Show Field Aliases, and click Fields View.

3. Scroll down in the Fields table, hold down the CTRL key, and select USER_Address and USER_ZIP_Code. These are input source values from the survey.

4. Drag the selected rows up and drop them after Match_addr near the top of the table.

5. Under the Fields tab, click Save and close the Fields table.

6. Sort the attribute table by Score Ascending.

7. Compare Match_addr (the matched address and zip code in the Streets reference feature class) with USER_Address and USER_ZIP_Code. As you scroll down the table, there are obvious errors and some good matches. Some Match_addr values do not have street numbers, such as 2nd St, 15045 (Score 79), while corresponding USER_Address does, 2321 2ND ST 15045. In such cases, the Streets reference data had some street segments with the correct name and **zip** code but without street numbers (some LFROMADD, LFTOADD, RFROMADD, and RTOADD values were missing) in the range needed for a match. The geocoding algorithm picks the center point on one such street segment, as an approximation. Such approximations are probably good enough for marketing purposes, so a score around 74 looks like a good minimum score. If, however, marketing needed more precision, with good matching addresses including street numbers, a score of 90 looks good. Use 74, and instead of regeocoding the survey data, next just set a file definition with the expression Score is greater than or equal to 74.

YOUR TURN

Considering a minimum score of 74, use Select by Attributes to find the number of matched, tied, and unmatched records for the total of 932 input records. You will find 896 (96%) matched. This is good geocoding performance.

Produce final geocoding results

1. Clear any selected records and close the AttendeesAlleghenyCounty_Geo1010 attribute table.

2. In Contents, right-click AttendeesAlleghenyCounty_Geo1010, click Properties > Definition Query, and add and apply a new definition query for Score is greater than or equal to 74.

3. In Contents, rename AttendeesAlleghenyCounty_GEO1010 AttendeesAlleghenyCounty, and symbolize it with red circle 3, size 5. There are obvious clusters of attendees as well as gaps with no attendees. This is useful spatial information for marketing the arts event.

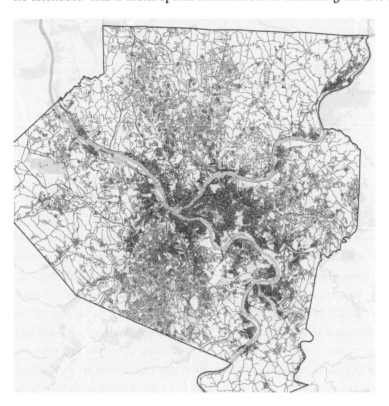

4. Save and close your project.

Assignments

This chapter has two assignments to complete that you can download from this book's resource page, go.esri.com/GISTforPro2.8Data:

*   **Assignment 8-1**: Geocode grocery stores in Allegheny County.
*   **Assignment 8-2**: Reverse geocode a random sample of points representing stops for Meals on Wheels kitchen siting.

PART 3

Applying advanced GIS technologies

CHAPTER 9

Spatial analysis

LEARNING GOALS

- Use buffers for proximity analysis.
- Use multiple-ring buffers to estimate a gravity model of demand versus distance from nearest facility.
- Estimate service areas of facilities using ArcGIS® Network Analyst.
- Optimally locate facilities using Network Analyst.
- Carry out cluster analysis to explore multidimensional data.

Introduction

Maps sometimes require more than a visualization of spatial data for users to answer questions and solve problems. The right data may be on the map, but analytical methods offer the best answers or solutions to a problem. This chapter covers four spatial analytical methods: buffers, service areas, facility location models, and clustering. The application areas include identifying illegal drug dealing within drug-free zones around schools, estimating a gravity model for the fraction of youths intending to use public swimming pools as a function of the distance to the nearest pool from their residences, determining which public swimming pools to keep open during a budget crisis, and determining spatial patterns of arrested persons for serious violent crimes.

This chapter introduces a new spatial data type, the network dataset, which is used for estimating travel distance or time on a street network. ArcGIS Online provides accurate network analysis with network datasets for much of the world. Because these services require purchase, ask your instructor if you have credits. For instructional purposes, this chapter uses approximate network datasets built from TIGER street centerlines that are free to use.

Tutorial 9-1: Buffers for proximity analysis

A buffer is a polygon surrounding map features of a feature class. As the analyst, you will specify a buffer's radius, and the Buffer tool will sweep the radius around each feature, remaining perpendicular to features, to create buffers. For points, buffers are circles; for lines, they're areas or rectangles with rounded end points; and for polygons, they're enlarged polygons with rounded vertices.

Generally, you use buffers to find what's near (proximate to) the features being buffered. For schools symbolized as points, 1,000-foot school buffers define drug-free zones. Any person convicted of dealing illicit drugs within such zones will receive a mandatory increase in jail sentence, which is intended to serve as a deterrent to selling drugs to children. You'll analyze drug-free zones in this tutorial.

Another example of spatial analysis with buffers is food deserts, which are areas that are more than a mile from the nearest grocery store in a city. Often, the persons living in food deserts are poor and from minority populations, and it's straightforward to analyze affected populations using selection by location. You will select the populations within buffers and analyze selected features—for example, using ArcGIS Pro to summarize data. Additionally, from ArcGIS Online, you will have the opportunity to explore this topic further. Assignment 9-2 addresses the analogous problem to food deserts of urgent health care deserts in Pittsburgh—areas more than a mile from the nearest urgent health care facilities. Assignment 9-4 is a response to food deserts, finding the best locations for additional farmers' markets in Washington, DC.

Sometimes, buffers are exactly the right tool. One such example is the drug-free zones for which federal and state laws prescribe a buffer radius, generally 1,000 feet, for school properties. Other times, buffers are less accurate but provide a quick estimate. One such case is geographic access by youths to public swimming pools in Pittsburgh, because users of the pools travel the street network to get to the pools. Pittsburgh has an irregular street network because of its rivers and hilly terrain, so even though some youths appear to be close to a pool on a map, they may have no direct route to the pool. In this case, you'll need a network model that uses travel distance on a street network dataset. Buffers estimated with street networks are called service areas, and you'll work with them in tutorial 9-3 to analyze public swimming pools in Pittsburgh.

Open the Tutorial 9-1 project

In this tutorial, you'll buffer Pittsburgh schools to find illicit drug dealing arrests within drug-free school zones—1,000-foot buffers of schools. Drug arrests often occur at the scene of drug dealing, so arrest locations, which are the same as illegal drug sales violations, are relevant for the analysis of this exercise.

1. **Open Tutorial9-1.aprx from Chapter9\Tutorials, and save it as** Tutorial9-1YourName.aprx.

2. **Use the Northside bookmark.** Pittsburgh (and any city) has so many schools that it seems most drug arrests, as shown here in the Northside of Pittsburgh, would likely fall into drug-free zones. For this tutorial, persons' names in the Drug Violations feature class are from a random-name generator and are not the actual names of arrested persons. Also, dates, exact addresses, and other data have been changed to protect privacy.

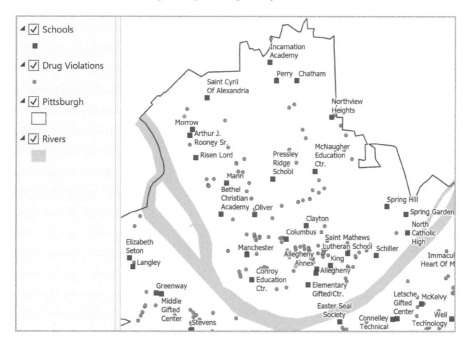

Run the Buffer tool

1. **Search for and open the** Buffer (Analysis Tools) **tool.**

2. **Type or make the following selections as shown.** The No Dissolve option creates a separate buffer for each point feature. The opposite, Dissolve, dissolves interior lines of overlapping buffers, merging them into a single buffer. A dissolved buffer gives you distance to the nearest facility (in this case, school).

*   **Input Features:** Schools
*   **Output Feature Class:** Schools_Buffer
*   **Distance:** 1000 **Feet**
*   **Dissolve Type: No Dissolve**

3. **Run the tool.**

4. **Take the color fill out of the buffers, move Drug Violations above Schools in the Contents pane, and turn off the school labels.** Looking at the map alone, you cannot readily determine whether drug-free zones deter drug dealing near schools in Pittsburgh. You'll analyze this issue in the next Your Turn assignment. Note the overlapping buffers in the figure. If you made the buffers with the Dissolve option instead, each set of overlapping buffers would be replaced with a single buffer consisting of the outline of each set.

Select drug violations within drug-free zones

1. On the Map tab, in the Selection group layer, click Select By Location.

2. For Input Features, select Drug Violations, and for Selecting Features, select Schools_Buffer.

3. Run the tool.

4. **Open the Drug Violations table, and see how many violations are selected.** Across the city, 222 of 651 drug violations are selected within drug-free zones, 33.8 percent of all drug violations.

5. Close the table, and clear the selection.

YOUR TURN

If drug violations occurred randomly in Pittsburgh, for any given area within Pittsburgh, you'd expect the fraction of such crimes to be the same as that area's fraction of Pittsburgh's area. Run the Buffer tool again with Schools as input, **Schools_Buffer_Dissolved** as output, a **1,000**-foot radius, and with the Dissolve all output features into a single feature option. Then divide the area of the resulting buffer by the area of Pittsburgh. Both areas are in square feet and so are large numbers. You can copy and paste them into your computer's built-in calculator or Microsoft Excel from the Shape_Area attributes of the two feature classes and do the division. You should get 398,645,239/1,627,099,663 = 0.245, or 24.5 percent, but earlier you found that a substantially higher fraction, 33.8 percent, of drug violations occurred in drug-free zones. Although not a definitive result, you should be suspicious that drug-free zones are not working in Pittsburgh. A better estimate than the one you just did uses Pittsburgh's area with human activity instead of all of Pittsburgh for the divisor. You'd have to subtract the area of rivers, cemeteries, steep hillsides, and so on, which for Pittsburgh is about 50 percent of its area. Then you'd expect 49 percent of drug arrests to be in the drug-free zone buffers, which is considerably larger than the 33.8 percent found. So perhaps the law is effective in reducing drug dealing near schools. Save your project.

Spatially join school buffers to drug violations

You can get a list of all drug violations within school buffers with the names of the schools included. If a drug violation is in more than one school buffer, you can get the names of all the schools. You'll use a spatial join of school buffers to drug violations to get this information, which could be passed along to police and the courts.

1. Open the Spatial Join tool.

2. Type or make the selections as follows, using your original Schools_Buffer.
   - **Target Features:** DrugViolations
   - **Join Features:** Schools_Buffer
   - **Output Feature Class:** DrugViolations Buffers Join
   - **Join Operation:** Join One To Many

Leave Keep All Target Features unchecked.

3.   **Run the Spatial Join tool, and close it when it finishes.**

Note: When you run the tool, if you get a warning message, you can ignore and dismiss it.

The message states that the value CHESTNUT ST & SPRING GARDEN AV caused a processing step to fail. The reason is that the value is too long. If you shorten it, say to CHESTNUT ST & SPRING GARDEN, then there is no warning message. Nevertheless, the corresponding record exists in the spatially joined output and with the value CHESTNUT ST & SPRING GARDEN AV included in the Address field. So, despite the warning, the join succeeded 100 percent.

4.   **Open the attribute table of the resulting spatial join.**

5.   **Sort the table by CCN (the crime ID) ascending.** Notice that the first three records are all for the same CCN number, 2015016991, so the corresponding drug offense is in three school buffers (see the Name attribute for the names of the three schools). This output is the result of the Join one to many option from the tool: one drug violation possibly in many buffers.

6.   **Save your project.**

Tutorial 9-2: Multiple-ring buffers

A multiple-ring buffer for a point looks like a bull's-eye target, with a center circle and rings extending out. You can configure the center circle and each ring to be separate polygons, thereby allowing you to select other features within given distance ranges from the buffered feature.

During a budget crisis, Pittsburgh officials permanently closed 16 of 32 public swimming pools. You'll estimate the number of youths, ages 5 to 17, living at different distances from the nearest swimming pool for all 32 pools versus the 16 that were kept open. Youths living within a half mile of the nearest open pool are considered to have good access to pools, whereas youths living from a half to one mile from the nearest pool are considered to have fair access. Youths living farther than one mile from the nearest pool are considered to have poor access (borrowing from the definition of food deserts). In tutorial 9-4, you'll make more precise access estimates on the basis of travel time across the street network of Pittsburgh from youth residences to the nearest pool.

Open the Tutorial 9-2 project

1. **Open Tutorial9-2.aprx, and save it as** Tutorial9-2YourName.aprx.

2. **Use the Pittsburgh bookmark.** The map has all 32 public pools (both open and closed) and block centroids symbolized with youth population, ages 5 to 17. First, you'll get the number of youths that had good access when all 32 pools were open.

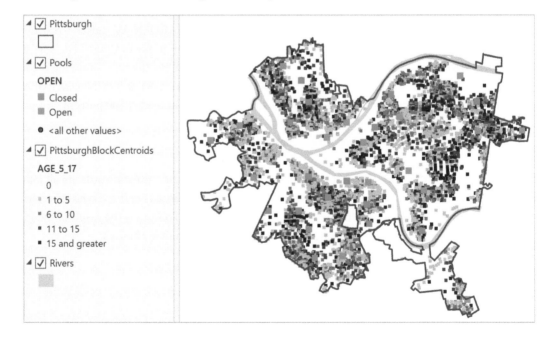

Create buffers

1. **Open the Multiple Ring Buffer tool, and type or make the selections as shown.**
   - **Input Features:** Pools
   - **Distance:** 0.5, 1
   - **Buffer Unit: Miles**

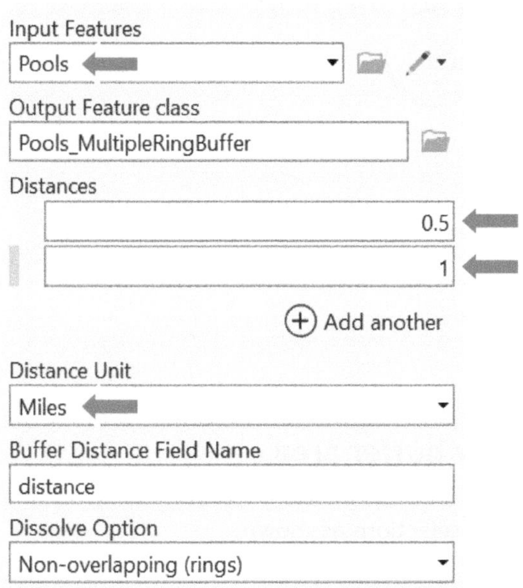

Input Features

| Pools ⬅ | ▼ |

Output Feature class

Pools_MultipleRingBuffer

Distances

| 0.5 ⬅ |
| 1 ⬅ |

⊕ Add another

Distance Unit

| Miles ⬅ | ▼ |

Buffer Distance Field Name

distance

Dissolve Option

| Non-overlapping (rings) | ▼ |

2. **Run the tool.**

3. **Symbolize the buffers to have no color fill with a size** 2 **dark-green outline.** Many Pittsburgh youths appear to have had fair-to-good pool access before the pool closings. Next, you'll get the corresponding statistics.

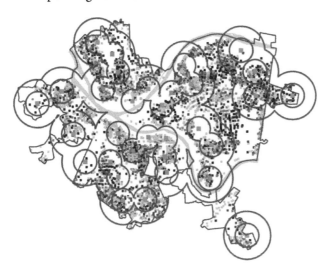

Use spatial overlay to get statistics by buffer area

1. **Open the Spatial Join tool, and type or make the selections as shown.**
    - **Target Feature:** Pools_MultipleRingBuffer
    - **Join Features:** PittsburghBlockCentroids
    - **Output Fields:** Age_5_17
    - **Merge Rule: Sum**

2. **Run the tool.**

3. **Open the resulting map layer's attribute table to see in the Age_5_17 attribute that 21,833 youths had good access, and 20,715 had fair access.** If you ran the Summarize tool to sum the Age_5_17 field of PittsburghBlockCentroids, you'd get 48,903 youths who live in the city. Then you can calculate that 44.6 percent had good access and 42.4 percent had fair access, for a total of 87.0 percent who had good or fair access. Those results seem to be good.

YOUR TURN

Next, find good or fair youth access for pools that remained open. Select open pools using Select By Attributes with the criterion Open Is Equal To 1. Create the same multiple-ring buffers for open pools. Clear your selection and perform a spatial join of the block centroids to the new buffers. This will give you new totals for good and fair access. You'll find that 10,726 youths have good access and 20,440 have fair access, compared with 21,833 who had good access and 20,715 who had fair access when all pools were open. Why do you suppose that fair access remained so high? Maybe it's the conversion of good access to fair access, or maybe it's because with all pools open, there tended to be access to more than one pool for a given residence, whereas with pools closed, there's still fair access but to only one pool. In any event, now only 63.7 percent of the youths have good or fair access compared with 87 percent with all pools open. Save your project.

Tutorial 9-3: Multiple-ring service areas for calibrating a gravity model

Service areas are like buffers but are based on travel over a network, usually a street network dataset. If a point, say for a retail store, has a five-minute service area constructed using Network Analysis tools in ArcGIS Pro, anyone residing in the service area has, at most, a five-minute trip to the store. If you have permission from your instructor to use an ArcGIS Online service that consumes credits or otherwise have access to your own ArcGIS Online credits, you could use an ArcGIS Online network service, which would be much more accurate than the free, TIGER-based network datasets that you'll use in this chapter. Nevertheless, you will use the PittsburghStreets_ND network dataset provided in Tutorial9-4.aprx so that your results match the tutorial results.

In this tutorial, you'll use service areas to estimate a gravity model of geography (also known as the spatial interaction model), which assumes that the farther apart two features are, the less attraction between them. The falloff in attraction with distance is often nonlinear and rapid, as in Newton's gravity model for physical objects where the denominator of attraction is distance squared. The application of this tutorial is a continuation of the pool case study, based on a random sample of youths owning pool tags (which allow admission to any Pittsburgh public pool). To scale the random sample up to the full population of youths with pool tags, you will need to multiply estimates by 11.3. With service areas, you could use distance or travel time to estimate a gravity model. Here, you'll use time (minutes).

This tutorial uses the following multiple-step workflow:

1. Select open pools from the feature layer of all pools (so that the service area of the next step is applied to only open pools).
2. Create service areas for a given set of travel times (1, 2, 3, 5, and 7 minutes) to produce non-overlapping service area rings. The result is the center area and four rings (0 to 1 minute, 1 to 2 minutes, and so on). These times were determined by trial and error to get good sample sizes in each ring and cover the catchment areas of pools.

3. Count the number of pool tags in each service area. The count is accomplished using a spatial join of the service areas and pool tag points. When scaled up by multiplying by 11.3, the result is an estimate of the number of youths owning pool tags who will use pools in each service area ring.

4. Sum up the population of youths ages 5–17 in each service area, using a spatial join of the service areas and block centroids. The result is the target total population for pool use of all youths in each service ring.

5. Calculate the average use rate for each service area as a percentage using Use Rate = 100 × 11.3 × (number of youths with pool tags in sample) / (total population of youths).

6. Plot the estimated use rate of each service area versus the average travel time in each service area in minutes from the nearest pool (0.5, 1.5, 2.5, 4.0, 6.0). Use rate is expected to decline rapidly with travel time.

Note for the technically advanced student: You can make more accurate estimates of the average travel times for each service area ring of step 6 using an origin destination (OD) cost matrix calculated using a Network Analyst tool (search for **OD cost matrix analysis** in ArcGIS Pro help). You can configure this matrix to record the shortest-path distance or time between each demand point (origin) and supply point (destination). In the current problem, block centroids are origins, and pools are destinations. Manipulations of the OD matrix, say in Microsoft Excel, can yield exact averages (or medians) of travel times for rings instead of assuming that the midpoints of service area rings are average travel times.

Open the Tutorial 9-3 project

1. **Open Tutorial9-3.aprx from Chapter3\Tutorials, and save it as** Tutorial9-3YourName.aprx.

2. **Use the Pittsburgh bookmark.** The map has all 32 public pools (both open and closed) and block centroids symbolized with youth population, ages 5 to 17, and pool tags. The map also has a network dataset for Pittsburgh, PittsburghStreets_ND, built from TIGER street centerlines. The network dataset is calibrated for drive time in a vehicle.

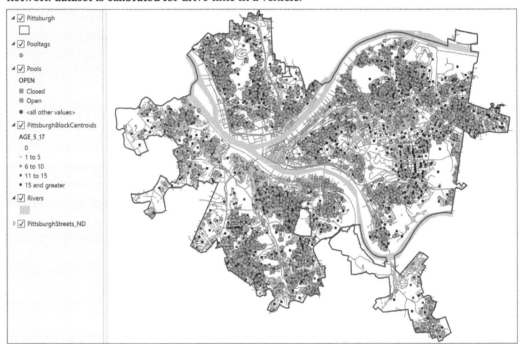

3. **Turn off Pooltags and PittsburghBlockCentroids.**

4. **Click Select By Attributes, complete its geoprocessing form as shown, and run the tool.** Later, when you add facilities (pools) to the network analysis, only open pools will be added.
   - **Input Rows:** Pools
   - **Expression:** Where Open Is Equal To 1

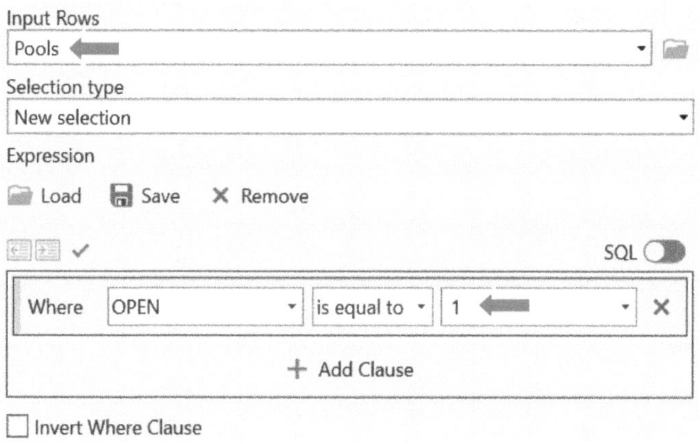

Input Rows

Pools ⬅

Selection type

New selection

Expression

📁 Load   💾 Save   ✕ Remove

✓   SQL ⬤

| Where | OPEN ▾ | is equal to ▾ | 1 ⬅ | ✕ |

+ Add Clause

☐ Invert Where Clause

Create multiple-ring service area polygons

1. **On the Analysis tab, in the Workflows group, click the Network Analysis button ▦ > Service Area.** If you don't see the Network Analysis button, make your ArcGIS Pro window wider. Because the Contents pane has a network dataset, PittsburghStreets_ND, the items on the Network Analysis list are enabled. ArcGIS Pro builds the Service Area group layer in the Contents pane and adds the Network Analyst service area context menu to the ribbon.

2. **Click the Network Analyst context tab > Import Facilities, and for Input Locations, select Pools. Leave Append to Existing Locations unchecked. Check Snap to Network. Run the tool.** All open pools become facilities because of their selection.

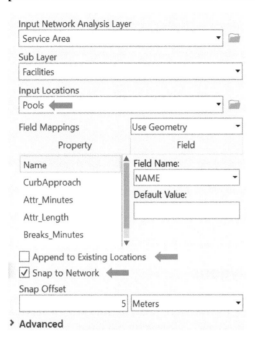

3. **In the Travel Settings group, change Direction to Towards Facilities.** Next, you'll change cutoffs for the service areas. ArcGIS Pro has detected that the street network, PittsburghStreets_ND, is set up for travel time in minutes and suggests 5, 10, and 15 minutes. You'll change these times next.

4. **For Cutoffs, type** 1, 2, 3, 5, 7.

5. **In the Output Geometry group, select Dissolve instead of Overlap, leaving Standard Precision and Rings selected.** By selecting Dissolve, your service areas will provide travel time to the nearest pool.

6. **On the Network Analyst contextual tab, in the Analysis group, click Run.**

7. **Symbolize the output polygons so that 1 is blue, 2 is bluish green, 3 is green, 5 is yellow, and 7 is orange. Also, turn on labeling for Pools, and move that layer above Service Area.** Suppose that a travel time of 1 minute or less is excellent, 1 to 2 minutes is very good, 2 to 3 minutes is good, 3 to 5 minutes is fair, and 5 minutes and higher is poor. Then according to estimated street travel time in a car, areas that are considered fair or poor are clearly visible in yellow and orange with two interesting results. First, the service areas are much different from circular pool buffers because Pittsburgh's streets have irregular patterns caused by the city's many hills and valleys. Jack Stack and Highland pools are good examples. Zoom in to their areas to see their streets. Second, you will see open pools close together, such as Cowley and Sue Murray, where perhaps one could be closed. Also, closed pools, such as Fowler, could be opened to provide better access. The analysis that you just performed was not available at the time of initial pool closings so these results would have been informative; however, city officials used criteria in addition to geographic access when closing pools, such as the condition of pools and historic attendance patterns. After the first year of closings, with location analysis and other information in hand, some closed pools were opened and others closed to improve overall pool access.

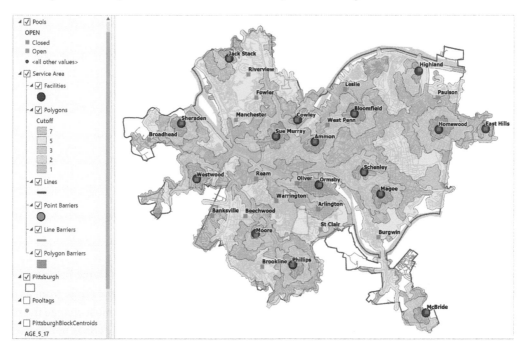

Next, you will count youths with pool tags and sum up youth population, all by service area polygons.

Spatially join service areas and pool tag owners

1. **Turn Pooltags on.**

2.   **In the Contents pane, right-click Polygons in the Service Area group, and click Joins and Relates > Spatial Join, type or make the selections as shown, run the tool, and close it when finished.** The Count Merge Rule counts pool tag records, and because there's only one youth per pool tag, counting records is needed.

-   **Join Features:** Pooltags
-   **Output Feature Class:** Join_Pooltags
-   **Merge Rule: First**
-   **Service Area\Polygons:** FacilityID

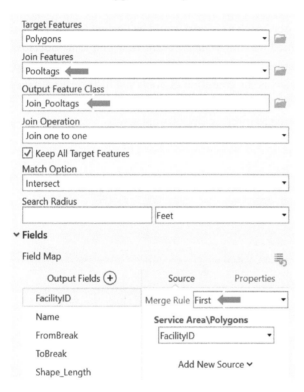

3.   **Open the resulting Join_Pooltags table and examine the Join_Count and Name fields to see that there are 374 sampled pool tag holders in the 0 to 1 minute travel-time range from all pools, 514 in the 1 to 2 range, and so on. Leave the table open.** Next, you must change the name of the Join_Count field so that the spatial join you run next on Join_Pooltags does not overwrite the current Join_Count values.

4.   **In the attribute table, click the Options button > Fields View, change the Field Name and Alias of Join_Count to** Join_Count_Pooltags, **save the changes, and close all tables.**

5. **In the Contents pane, right-click the Join_Pooltags layer, click Joins and Relates > Spatial Join, type or make selections as shown, and run the tool.** Note that you can specify the Merge rule for each field in the Output Fields panel as shown in the figure.

    - **Join Features:** PittsburghBlockCentroids
    - **Output Feature Class:** Join Pooltags Blocks
    - **Output Fields:** Age_5_17
    - **Merge Rule: Sum**

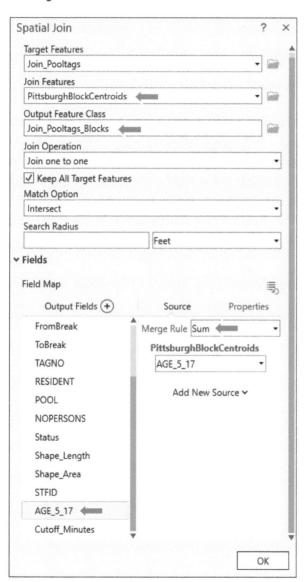

Spatially join service areas and youths

Next are the final steps of the workflow—calculating and plotting use rate.

1. **Open the attribute table of Join_Pooltags_Blocks.**

2. **Open Fields View of Join_Pooltags_Blocks, and create two fields:** AverageTime **and** UseRate, **both with the Float data type.**

3. **Save the results, and close the Fields View table.** AverageTime stores the midpoint of each service area ring—for example, 0.5 for the 0–1 ring.

4. **Move the Name column to just before the AverageTime field.**

5. **Type the following values for AverageTime, referring to Name to correspond to travel time ranges:** 6, 4, 2.5, 1.5, **and** 0.5. The final task is to compute estimated use rates for each service area ring = $100 \times 11.3 \times$!Join_Count_Pooltags!/!Age_5_17!, where 100 makes the result a percentage, 11.3 is the scale-up factor from the random sample to the population, Join_Count_Pooltags is the sample number of youth pool tag holders in each service area ring, and Age_5_17 has the total (sum) of youths in each ring.

Name	AverageTime
5 - 7	6
3 - 5	4
2 - 3	2.5
1 - 2	1.5
0 - 1	0.5

6. **On the Edit tab, save your edits.**

7. **Right-click UseRate, click Calculate Field, and type or make selections as shown.**
    - **Field Name (Existing or New):** UseRate
    - **Expression:** 100 * 11.3 * !Join_Count_Pooltags! / !AGE_5_17!

Input Table

| Join_Pooltags_Blocks | ▾ | 🗁 |

Field Name (Existing or New)

| UseRate | ▾ |

Expression Type

| Python 3 | ▾ |

Expression

Fields ▽   Helpers ▽

Fields	Helpers
TAGNO	.as_integer_ratio()
RESIDENT	.capitalize()
POOL	.center()
NOPERSONS	.conjugate()
Status	.count()
STFID	.decode()
AGE_5_17	.denominator()
Cutoff Minutes	

| Insert Values | ▾ | *   /   +   -   = |

=

| 100 * 11.3 * !Join_Count_Pooltags! / !AGE_5_17! |

8. **Run the tool.** As expected, use rate declines quickly with average time from the nearest pool: 91.39 percent of youths within a mile of the nearest pool have pool tags, but the rate drops to 54.42 percent in the 1 to 2 mile ring, and all the way down to 25.86 percent in the 5 to 7 minute ring.

Name	AverageTime	UseRate
5 - 7	6	25.86141
3 - 5	4	37.24302
2 - 3	2.5	44.22211
1 - 2	1.5	54.42466
0 - 1	0.5	91.39706

Make a scatter plot

In ArcGIS Pro, you can make a scatter plot of Use Rate versus Average Time to visualize the estimated gravity model data points.

1. **In the Contents pane, select Join_Pooltags_Blocks.**

2.   **On the Feature Layer's contextual Data tab, in the Visualize group, click Create Chart > Scatter Plot.** If you don't see the Create Chart button, make your ArcGIS Pro window wider.

3.   **In the Chart pane, for X-Axis Number, select AverageTime; for Y-Axis Number, select UseRate; and close the Chart pane.** The gravity model curve falls rapidly in a nonlinear way as expected. Note that you can change the title and axis labels of the chart using the General tab of the Chart pane. Automatically included in the chart is a straight-line regression model, which is not an appropriate form for the evidently nonlinear relationship.

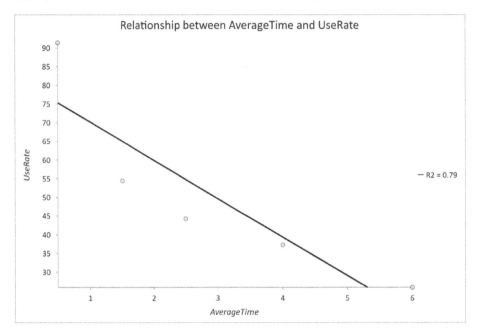

4.   **Save your project.**

Tutorial 9-4: Facility location

Suppose that you are an analyst for an organization that owns several facilities in a city, and you are asked to find the best locations for new facilities. The classic problem of this kind is to locate facilities for a chain of retail stores in an urban area, but other examples are federally qualified health centers (FQHCs) as studied in chapter 1 and the public swimming pools earlier in this chapter. In the most general case, your organization has existing facilities that will remain open, a set of competitor facilities, and a set of potential new locations from which you want to determine a best subset of a specified size. Another case is where a subset of existing facilities must be closed, as with the Pittsburgh swimming pools, for which you want to determine the appropriate 16 of 32 facilities to close. Yet another case is where there are no existing facilities, and you want to locate one new facility.

In ArcGIS Pro, the Location-Allocation model in the Network Analysis collection of models handles these sorts of facility location problems. Inputs are a network dataset, locations of facility types (existing, competitors, and new potential sites), demand points, and a gravity model. Demand is represented by polygon centroids—blocks, block groups, tracts, zip codes, and so on—for which you have data on the target population, generally from the US Census Bureau, such as youth population. Resistance to flow in the network, called impedance, is represented by a gravity model and can be distance or time traveled along shortest paths to facilities. Several optimization models are available within the Location-Allocation model. You'll use the Maximize Attendance model, which includes a gravity model (for which you supply parameter values) and a network optimization model that selects a specified number of new facility locations that maximize attendance.

In this tutorial, you'll run a model to choose the best 16 out of 32 swimming pools to keep open using geographic access (distance from the nearest pool) as the criterion.

Open the Tutorial 9-4 project

1. **Open Tutorial9-4.aprx from Chapter9\Tutorials, and save it as** Tutorial9-4YourName.aprx.

2. **Use the Pittsburgh bookmark.** This map has the 32 pools, block centroids with youth population, and PittsburghStreets_ND network dataset. Also included is the PittsburghStreets street center-line layer symbolized using a tan color for visualization. Note that the PittsburghBlockCentroids table has an attribute, Cutoff_Minutes, which has value 10, meaning that any block centroid farther than 10 minutes travel time to the nearest pool under consideration will not be served and therefore has none of its youth going to a pool.

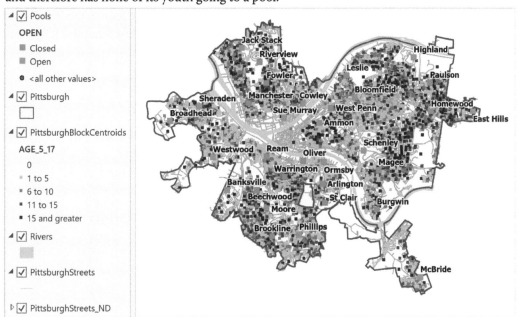

Calibrate the Location-Allocation model

The Location-Allocation model is straightforward to use, but first you must fit a gravity model to the five points of the scatter plot at the end of tutorial 9-3. In other words, tutorial 9-3 produced data points, but the Location-Allocation model needs a gravity-model function fitted to those points as input. Available in the Location-Allocation model are three functional forms for a gravity model: linear, exponential, and power. The power form does not work well for cases with short travel times, such as for the swimming pools (a few minutes), so it's not discussed further here. The linear form is based on an impedance cutoff (10 minutes for the swimming pools). It estimates that 100 percent of the target population that lives at a pool uses it (of course, no one lives at a pool, but some are nearby, and nearly 100 percent of the population uses the pool), 75 percent uses the pool at a quarter of the cutoff (2.5 minutes), 50 percent at half the cutoff time (5 minutes), 25 percent uses it at three quarters of the cutoff (7.5 minutes), and 0 percent at the cutoff or beyond—(10 minutes or higher). If C = cutoff in minutes and T = impedance in minutes, then as a percentage, Use Rate = $100 \times (1 - T/C)$ for T ranges between 0 and C, and is 0 otherwise.

Exponential is the most applicable gravity model for the swimming pool case, because it declines rapidly as travel time increases, and it generalizes to other cases very well. The Microsoft Excel worksheet, Exponential.xlsx, available in Chapter9\Data, provides a method of fitting the exponential model to the results of tutorial 9-3.

1. **Open Exponential.xlsx from Chapter9\Data.** Beta, which currently has the value 0.25, is the one parameter that you must choose (optimize) in the worksheet and provide in the Location-Allocation model. Impedances are average travel times in minutes. The Gravity Model Use Rate column has the estimates between 9 and 3 for its multiring service areas, expressed here as fractions instead of percentages. Cost is an exponential function, $e^{\beta T}$, used in the network optimization model to represent system impedance, and the model uses Pro Use Rate, $e^{-\beta T}$, as the gravity model. Your task is to vary beta until you get a good fit of the Pro Use Rate to the Gravity Model Use Rate. Average Absolute Error is a guide for choosing beta; you must minimize it by trial and error.

	A	B	C	D	E
1	Beta =	0.25 ⬅			
2					
3	T = Impedance	Gravity Model	Cost =	Pro Use Rate =	Absolute Error
4	(minutes)	Use Rate	Exponential	1/Exponential	
5	0.5	0.913	1.133	0.883	0.030
6	1.5	0.544	1.455	0.687	0.143
7	2.5	0.442	1.868	0.535	0.093
8	4	0.372	2.718	0.368	0.004
9	6	0.259	4.481	0.223	0.036
10				Average Absolute Error	0.061

2. **Enter the following beta values, one at a time, into the worksheet:** 0.10, 0.15, 0.20, 0.25, 0.30, and 0.35, **and note resulting average absolute error values.** Next are the results, which are also saved to the worksheet in rows 13 through 19. Beta = 0.25 is the best fit of the values tried. Note that if you are comfortable with Microsoft Excel, you can modify this worksheet for other case studies or projects by adding (copying) rows and changing impedances.

Beta	Average Absolute Error
0.10	0.255
0.15	0.167
0.20	0.099
0.25	0.061 ⬅
0.30	0.069
0.35	0.081

3. Save and close the worksheet.

Configure and run the Location-Allocation model

1. **On the Analysis tab, in the Workflows group, click Network Analysis > Location-Allocation.** ArcGIS Pro creates the Location-Allocation group layer in the Contents pane.

2. **Click the Network-Analyst contextual tab.** The horizontal toolbar for this model opens. If this toolbar ever closes and you want to reopen it, click the Location-Allocation group layer in the Contents pane.

3. **In the Input Data group, click Import Facilities, make the selections as shown, and run the tool.** You can run this tool more than once to load different kinds of facilities. Notice that the default value for FacilityType under Field Mappings is Candidate, which is correct for this model run. Other values are Required, Competitor, and Chosen.
   - **Input Locations:** Pools
   - **Property:** Facilitytype
   - **Default Value:** Candidate

Leave Append to Existing Locations unchecked. Check Snap to Network.

4.   Click Import Demand Points, and make the selections as shown. For Input Locations, select PittsburghBlockCentroids. Under Field Mappings Property, select Weight and set the Field Name to AGE_5_17 (this is the target population), and for the Cutoff_Minutes property, select the Cutoff_Minutes field. Check Append to Existing Locations. Leave Snap to Network unchecked. Run the tool.

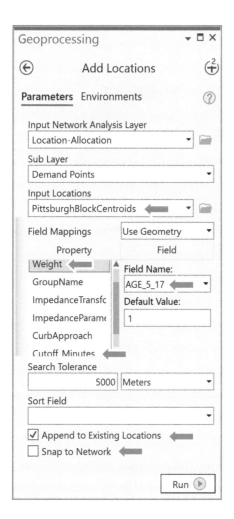

It takes a minute or so for ArcGIS Pro to load the 7,493 demand points.

5. **In the Travel Settings group, select Towards Facilities, and type** 16 **for Facilities.**

6. **In the Problem Type group, click Type > Maximize Attendance.**

7. **In the Problem Type group, select Exponential for f(cost, β), type** 0.25 **for β, and press Tab.**

8. **Run the model.**

9. **In the Contents pane and its Location-Allocation group layer, clear Demand Points, and resymbolize Lines to have a** 0.5 **line width and color of your choice.** The lines connecting pools and served demand points constitute a spider map. This sort of map is used for visualization; however, the actual lines connecting pools and block centroids follow the shortest paths along the street network.

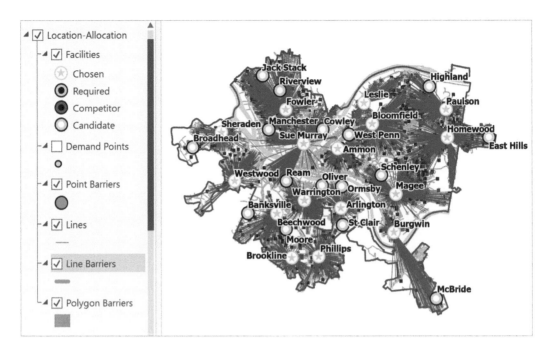

Analyze the optimal solution

Officials had closed half (eight) of the optimal pools. Remember, however, that officials had criteria in addition to geographic access for selecting pools to close or keep open and didn't have GIS-based location analysis.

1. **Open and sort the Facilities attribute table as descending using the DemandWeight column.** DemandWeight has the number of youths allocated by the Location-Allocation model to each pool. However, only about 50 percent of Pittsburgh's youths had pool tags in the study year, so a better estimate of the number of users for each pool is to divide the DemandWeight values by 2. Then, for example, you'd expect 3,317/2 = 1,658 users of the Homewood pool.

Name	DemandWeight ▾
Homewood	3317.132472
Fowler	3305.723629
Warrington	2430.780413
Bloomfield	2346.782229
Magee	2010.08547
Paulson	1872.731705
Ammon	1692.61658
Sue Murray	1528.067438
Arlington	1266.629243
Brookline	1240.856129
Sheraden	1229.166622
Phillips	1213.203297
Leslie	1187.772994
Beechwood	1119.83618
Westwood	946.977381
Burgwin	899.124667

2. **Open the Pools attribute table. MaxLoad is the capacity of a pool.** The Homewood pool has a capacity of 455, whereas the model estimate is $3{,}306/2 = 1{,}653$—more than 3.5 times its capacity. Of course, not every potential user of the pool will show up at once. Some days, youths would have to wait or be turned away from the pool, but it seems that attendance would be good.

3. **Close the Pools table.**

4. **In the Facilities attribute table, use the Summarize tool to sum DemandWeight.** The total is 27,607 estimated pool users. Dividing by 2, the estimate is that 13,804 youths would use the 16 optimal pools.

5. **Close the Facilities table.**

6. **In the Contents pane, clear PittsburghStreets and PittsburghStreets_ND, zoom in to the south-east corner of Pittsburgh that has the long lines from the Burgwin pool to demand points near the McBride pool, and select the longest line you can find.** In the resulting pop-up window, Total_Minutes, 7.2, has the estimated travel time from Burgwin to the farthest point on the map, within the 10-minute cutoff. If you have trouble selecting the longest line, select a few lines in a neighborhood far away from a pool, open the attribute table, switch to Show Selected Records and you will get a sense of the commute time for youths living in that area.

7. **Save your project.**

YOUR TURN

Remove the Location-Allocation group layer from the Contents pane, and set up a new Location-Allocation model to estimate use of pools that officials left open. First, select open pools with the clause Open Is Equal To 1. Then start a new Location-Allocation model, and import Pools as Input Locations with FacilityType (under Property) having a default value of Required. Turn off Append To Existing Locations, and turn on Snap To Network. Then import Demand Points just as you did in the previous steps (Weight = Age_5_17, and Cutoff_Minutes = Cutoff_Minutes). Select Towards Facilities, and type **16 facilities**. Set the problem type to Maximize Attendance and the function to Exponential with 0.25 for beta. Run the model. Sum DemandWeight for Facilities to find 25,044 youths estimated to use the open pools. Dividing by 2, you get the number 12,522 as the best estimate of pool users, compared with 13,804 from the optimal solution, only 1,282 fewer (9.3 percent). Ultimately, although officials

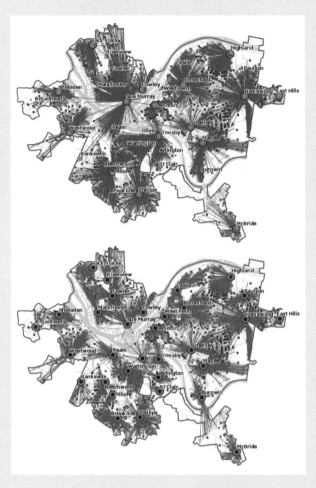

chose a much different set of pools to keep open compared with the model-based optimum, not much was given up in terms of potential users. That's good news for youths living in Pittsburgh from a policy point of view.

If you have time, make one more model run, this time using all 32 pools. You'll find that the model estimates 30,218 users compared with 27,607 for the optimal solution and 25,044 for the remaining 16 pools. Why do all 32 pools produce such a small gain over 16 pools? The answer is probably because there were too many pools competing for users when there were 32 pools. So it appears that officials did not make a bad decision, if the network dataset is accurate enough and travel to pools is primarily by driving on the road network.

Tutorial 9-5: Data clustering

The goal of data mining is exploration, to find hidden structure in large and complex datasets that has some interest or value. Data clustering, a branch of data mining, finds clusters of data points that are close to each other but distant from points of other clusters. If your data points were two-dimensional and graphed as a scatter plot, it would be easy for you to draw boundaries around points and call them clusters. You'd do as well as cluster methods. The problem is that when the points lie in more than 3D space, you can't see the clusters anymore. In this tutorial, you'll cluster crimes using three attributes or dimensions, including severity of crimes, plus the age and gender of arrested persons.

A limitation of clustering, however done, is that there is no way of knowing true clusters in real data to compare with what an algorithm determines are clusters. You take what clustering methods provide, and if you get information or ideas from cluster patterns in your data, you can confirm them or determine their predictability using additional models or other methods. Clustering is purely an exploratory method. There are no right or wrong clusters, only more or less informative ones.

This tutorial uses k-means clustering, a simple method available in the Multivariate Clustering tool. K-means partitions a dataset with n observations and p variables into k < n clusters. In the tutorial, you'll use a dataset with n = 303 observations, p = 3 variables for clustering, and k = 5 clusters. K-means is a heuristic algorithm as are all clustering methods: it's a set of repeated steps that produces good, if not optimal, results. For this tutorial's data, the algorithm starts with a procedure that selects five 3D observations called seeds as initial centroids for clusters. Then each observation is assigned to its nearest centroid, based on Euclidean (straight line) distance in the 3D cluster variable space. Next, a new centroid is calculated for each cluster, and all observations are reassigned to the nearest centroid. These steps are repeated until the cluster centroids do not move appreciably. So k-means clustering is just a common-sense method.

K-means assumes that all attributes are equally important for clustering because it uses distance between numerical points as its basis. To meet this assumption, it's important that you scale all input attributes to similar magnitudes and ranges. Generally, you can use standardized variables (for each variable, subtract its mean and divide by its standard deviation) to accomplish scale parity, but other ways of rescaling are acceptable too. It is the range, or relative distances between observations, that's important in k-means clustering. The data used in this tutorial includes numerical (interval or ratio), ordinal, and nominal class data for classification; whereas, strictly speaking, k-means clustering is intended for numerical data because of its use of distance in cluster variable space. Nevertheless, with rescaling it's possible to get informative clusters when including nonnumerical data. So next, the discussion turns to the specific case at hand and how to rescale attributes.

The data used in this tutorial is serious violent crimes from a summer in Pittsburgh, with the data mapped as points. The crimes are ranked by seriousness using FBI hierarchy numbers (1 = murder, 2 = rape, 3 = robbery, and 4 = aggravated assault) with, of course, murder being the most serious. Clearly, the nature of crimes should be important for their clustering. So the first assumption you have to make is that the distance between crime types, such as 3 between 1 for murder and 4 for aggravated assault (attempted or actual serious bodily harm to another person), is meaningful for clustering purposes. The criminal justice system agrees on the order, and for clustering you can accept the distances or change them using your judgment. You'll leave them as listed here.

Next, consider the single numerical attribute that is available: age of arrested person. Crime is generally committed by young adults, tapering off with older ages. For the serious violent crimes studied here, age varies between 13 and 65 (range of 42) with a mean of 29.7. Together with crime seriousness, age would dominate clustering because of its greater range. The remedy is to standardize age. Then it varies between −2.3 to 2.7 with a range of 5, whereas crime seriousness has a range of 3. So both attributes are fairly equal in determining clusters.

Finally, there is a nominal attribute, gender (male or female). You can encode this attribute as a binary attribute: 0 = male, 1 = female. As a binary indicator variable, gender has a mean, which is the fraction of arrestees who are female. This variable as encoded would have perhaps a lesser role than the previous two, but not by that much. If you wanted to increase the importance of the binary variables for clustering, you could encode it as (0, 2) or (0, 4) indicators. You'll leave it as a (0, 1) variable, which makes interpretation of clustering results easier.

One last point is that you must choose the number of clusters instead of having k-means clustering find an optimal number for you. That's the case for most clustering methods. For the crime data, experimentation with three to six clusters resulted in five clusters being the most informative, so you'll run with five clusters.

In summary, each observation is a 3D vector (crime, standardized age, gender); for example, (1, −0.364, 0) is a murder with an arrested 25-year-old (standardized age 25 is −0.364) male. The clusters found by k-means exist in the 3D space in which the observations lie. Each cluster is characterized by its centroid with the corresponding means of each cluster variable.

Open a map project

1. **Open Tutorial9-5.aprx from Chapter3\Tutorials, and save it as** Tutorial9-5YourName.aprx.

2. **Use the Pittsburgh bookmark.** The map shows the spatial distribution of serious violent crimes by crime type within police zones. Each zone has a commander, station, officers, and staff. Also shown are the poverty areas from chapter 1. Generally, there is a positive correlation between poverty and crime. Police do not record any measures of poverty for arrested persons, such as annual income, so that one cannot readily include poverty as a clustering variable. The map adds poverty as an additional variable for interpretation of clustering results.

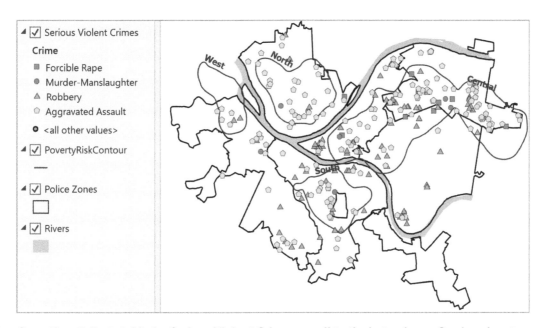

3. **Open the attribute table for Serious Violent Crimes, scroll to the last column, Seed, and sort the data descending.** The five records with Seed value of 1, found in a previous run of the Multivariate Clustering algorithm, are used as initial cluster centroids. The Hierarchy attribute is the FBI code for crime types and one of the three cluster attributes. ArrAge is the age of the arrested person and is the second cluster variable. Multivariate clustering automatically standardizes numerical variables, so ArrAgeStnd, which is ArrAge standardized, is not needed. ArrGender is the remaining cluster variable. Notice that there is a good deal of variation in the cluster variables of the seed records.

Crime	ArrAge	DateOccur	DayOfWeek	ArrAgeStnd	ArrGender	Seed ▾
Robbery	14	6/1/2015	Monday	-1.299793	0	1
Murder-Manslaughter	57	6/12/2015	Friday	2.226864	1	1
Robbery	39	6/19/2015	Friday	0.750589	0	1
Aggravated Assault	47	7/22/2015	Wednesday	1.406711	0	1
Aggravated Assault	14	8/9/2015	Sunday	-1.299793	0	1
Aggravated Assault	39	6/1/2015	Monday	0.750589	1	0

4. **Close the table.**

Run the clustering algorithm

The k-means algorithm is in the geoprocessing tool Multivariate Clustering, which you'll run next.

1. **Open the Multivariate Clustering tool.**

2. **Carefully fill out everything on the tool form as shown. Use the Tab key to move to a new field; do not use the Enter key because doing so may run the tool prematurely. Note that if you were**

running this tool for your own data or in an assignment, you'd use the Optimized Seed Locations instead of User Defined Seed Locations for Initialization Method.

- **Input Features:** Serious Violent Crimes
- **Output Features:** SeriousViolentCrimes_Clusters
- **Analysis Fields:** Hierarchy, ArrAge, ArrGender
- **Clustering Method: K-means**
- **Initialization Method: User-Defined Seed Locations**
- **Initialization Field:** Seed

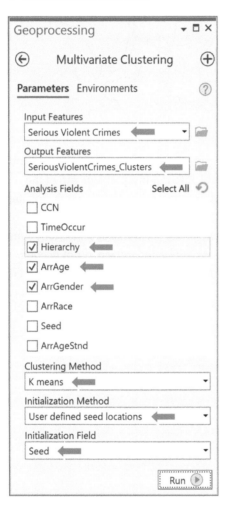

3. **Run the tool.** When finished, ArcGIS Pro adds the SeriousViolentCrimes_Clusters layer, which has the five identified clusters, to the Contents pane. The warning message says that the tool was not able to read 38 of the 303 crime records, but those are only records that did not geocode successfully in a data preparation step. Those 38 records are in the attribute table but not on the map, and they are not a problem in terms of successfully computing clusters for the 265 geocoded records.

4. **In Contents, turn off Serious Violent Crimes.**

YOUR TURN

As an aid in analyzing the resulting clusters, next run the Summary Statistics tool and calculate means for clustered variables by cluster. The statistic type is Mean for Statistics fields Hierarchy, ArrAge, and ArrGender.

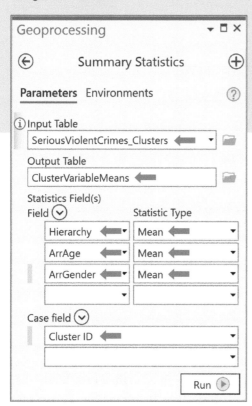

Cluster ID	FREQUENCY	MEAN_Hierarchy	MEAN_ArrAge	MEAN_ArrGender
1	75	2.64	22.706667	0
2	2	1	51	1
3	63	3.603175	45.984127	0
4	70	4	24.428571	0
5	55	3.818182	29.418182	1

Interpret clusters

The following table has the rows and values of the Summary Statistics table to which labels have been added for ranges of cluster variable means.

ClusterID	Frequency	Crime	Age	Gender
1 Young males moderate	75	2.6 moderate	23 young	0 male
2 Middle-aged females highest	2	1.0 highest	51 middle age	1 female
3 Middle-aged males lowest	63	3.6 low	46 middle age	0 male
4 Young males lowest	70	4.0 lowest	24 young	0 male
5 Young females lowest	55	3.8 lowest	29 young	1 female

These results have moderately interesting patterns and one anomalous group, 2. With a group size of only 2 crimes for group 2, you can't rely on the result that women in their early 50s or thereabouts are especially dangerous murderers. That's not a pattern likely to be repeated. Group 1 is young males committing a range of serious violent crimes. Group 3 is middle-aged criminals committing crimes toward the lesser end of serious violent crimes, mostly aggravated assaults (FBI hierarchy 4). Group 4, young males, is committing aggravated assaults. Finally, group 5, young females, mostly commit aggravated assaults. Next see if there are any spatial patterns for these groups.

1. **Open Symbology for SeriousViolentCrimes_Clusters, and relabel the groups as follows:**
    1   Young males middle
    2   Middle-aged females highest
    3   Middle-aged males lowest
    4   Young males lowest
    5   Young females lowest

2. **Keep the same colors but change the point symbols for the three young groups to Square 1 size 5.** The cluster results that were judged moderately interesting earlier get more interesting when mapped. The serious violent crimes in Pittsburgh's central business district (arrow in middle of map) are predominantly by middle-aged criminals. Crime patterns in central business districts

of cities are often unique in part because most persons present in those districts travel to them from their residences over some distances. Youths tend to commit crimes near where they live; whereas older criminals, who have higher mobility, travel to the central business district to commit crimes. Most crimes in poverty areas are by young persons, and the distance-to-crime theory of criminology states that these persons tend to commit crimes near where they live, within a mile or so. Somewhat surprising is the high percentage, 21 percent, of the serious violent crimes in the data being aggravated assaults committed by females. Those crimes are highly scattered across Pittsburgh except the southwest police zone, which only shows one such crime. Of interest in that zone is the group of serious violent crimes committed by youths in an area not considered to be a poverty area (indicated by the lower-left arrow).

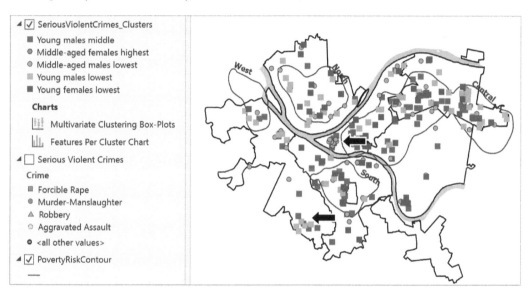

3. Save your project.

Assignments

This chapter has five assignments to complete that you can download from ArcGIS Online, at go.esri.com/GISTforPro2.8Data:

- **Assignment 9-1**: Study California cities affected by earthquakes.
- **Assignment 9-2**: Analyze geographic access to federally qualified health centers.
- **Assignment 9-3**: Analyze visits to the Phillips public pool in Pittsburgh.
- **Assignment 9-4**: Locate new farmers' markets in Washington, DC.
- **Assignment 9-5**: Carry out a cluster analysis of tornadoes.

CHAPTER 10

Raster GIS

LEARNING GOALS

- Extract and symbolize raster maps.
- Create hillshade maps.
- Smooth point data with kernel density smoothing.
- Build a raster-based risk index.
- Build a model for automatically creating risk indices.

Introduction

Thus far, this book has been devoted to vector feature classes (points, lines, and polygons), except for displaying the raster basemaps available in ArcGIS Pro, plus an occasional raster map layer. Vector feature classes are useful for discrete features (such as streetlight locations, streets, and city boundaries). Raster map layers are best suited for continuous features (for example, satellite images of Earth, topography, and precipitation). You can also use raster map layers to display large numbers of vector features (for example, all blocks in a city, all counties in the United States). In some cases, you have so many vector features that the features of choropleth and other maps are too small to render clearly. In such cases, as you will see in this chapter, you can transform vector feature classes into raster datasets that you can easily visualize. This chapter first presents some background on raster maps before discussing exactly what you'll be doing next.

Raster dataset is the generic name for a cell-based map layer stored on a disk in a raster data format. Esri supports more than 70 raster dataset formats, including familiar image formats such as TIFF and JPEG, as well as GIS-specific formats such as Esri Grid. You can import raster datasets into file geodatabases.

All raster datasets are arrays of cells (or pixels)—each with a value and location and rendered with a color when they are mapped. Of course, as with any other digital image, the pixels are so small at intended viewing scales that they are not individually distinguishable. The coordinates for a raster dataset are the same kind used for vector maps (see chapter 5).

All raster datasets have at least one band of values. A band is comparable to an attribute for vector map layers but stores the values of a single attribute in an array. The values can be positive or negative integers or floating-point numbers. You can use integer values for categories (codes), which must have a layer file with descriptions and colors (for example, 1 = Agriculture, brown; 2 = Forest, green). Raster dataset values can also be floating-point numbers representing magnitudes (for example, temperature and slope steepness of terrain).

Color capture and representation in raster datasets is an important topic. Color in the visible range is captured by satellites in three bands (for example, red, blue, and green) that mix to produce any color. Color in many raster datasets, however, is often represented in one band using a color map, in which each color is given a code (integer value). *Color depth* is the term given to the number of bits for code length (on/off switches in data storage) used to represent colors. True color uses 24 bits per pixel and can represent more than 16 million colors (the human eye can distinguish about 10 million colors).

The spatial resolution of a raster dataset is the length of one side of a square pixel. If a pixel is 1 meter on a side, it has 1-meter spatial resolution (which is a high resolution leading to high-quality maps). The US Geological Survey provides imagery for urban areas in the United States at this resolution or higher with images that can be zoomed in to small parts of neighborhoods (for example, with driveways, swimming pools, and tennis courts being clearly shown when displayed). The current Landsat 7 and 8 satellites that together image the entire Earth every eight days have a resolution of 30 meters for most of their eight bands, which is good for viewing areas as small as neighborhoods of a city.

File sizes for raster datasets can be large, requiring large amounts of disk space for storage and taking potentially long times to process and display on a computer screen. Raster GIS uses several mechanisms to reduce storage and processing time, including data compression and pyramids. Pyramids provide additional raster layers with larger spatial resolutions for zoomed-out viewing that take less time to display than the original layer. A mosaic dataset is a data catalog for storing, managing, viewing, and querying collections of raster datasets, often forming a continuous map when viewed. Although such a mosaic dataset is viewed as a single mosaicked image, you also have access to each dataset in the collection. A mosaic dataset also can store raster datasets of the same area for different times, with viewing of any time period's map and comparison of different time periods.

Tutorial 10-1: Process raster datasets

The ArcGIS Pro project that you will open has single-band raster datasets for land use and elevation, downloaded from the US Geological Survey's website. You'll extract raster datasets for Pittsburgh from each original dataset that has extents larger than Pittsburgh's. Because raster datasets are rectangular, you'll display layers using Pittsburgh's boundary as a mask: pixels in Pittsburgh's rectangular extent but outside the city's boundary will still exist but will be given no color, while those within Pittsburgh will have assigned colors. Finally, you'll use the elevation layer to produce a hillshade layer, which is a shaded relief rendering of topography created using an artificial sun to add illumination and shadows.

Open the Tutorial 10-1 project

1. **Open Tutorial10-1 from the Chapter10\Tutorials folder, and save it as** Tutorial10-1YourName.

2. **Use the Raster Datasets bookmark.** National Elevation Dataset (NED) is elevation data, and it will look like topography after you create a hillshade for it. LandUse_Pgh needs symbology for interpretation.

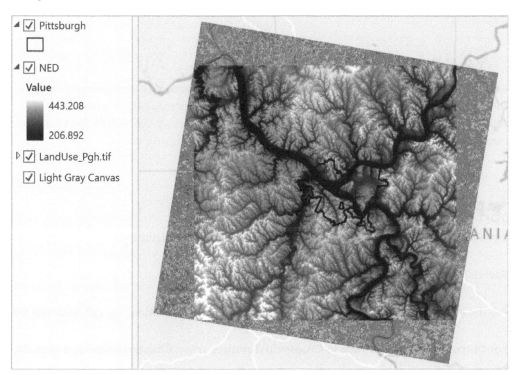

Examine raster dataset properties

Raster datasets have considerable metadata that you can read as properties.

1. **In the Contents pane, right-click NED, and click Properties > Source.** The dataset is in the Chapter10.gdb file geodatabase, where the dataset was imported from its original raster format. The elevation units are in feet above sea level.

2. **Click Raster Information.** The raster dataset has one band with 1,984 rows and 2,106 columns, for a total of 4,178,304 cells. The cells have roughly 90-foot (30-meter) resolution. The raster has pyramids for speedy display when zooming out.

3. **Scroll down, and click Statistics.** The average elevation in the raster dataset is 323.7 feet, with a minimum of 206.9 feet and a maximum of 443.2 feet above sea level.

4. **Click Extent.** The four values provided are state plane coordinates in feet that you can use to specify the four corners of the rectangular extent of the raster dataset.

5. **Click Spatial Reference.** Here, you see that the coordinates are in state plane feet for southern Pennsylvania, which uses a Lambert Conformal Conic projection tuned for southern Pennsylvania.

6. **Click Cancel.**

YOUR TURN

Review the properties of LandUse_Pgh. Note that the raster format is TIFF, a common image format. Its resolution is 30 meters, with a single band and a color map (which is available in a separate layer file that you'll use in an exercise that follows). It has a projection for the Lower 48 states (Albers Equal Area) that distorts direction (which explains why it's tilted).

Import a raster dataset into a file geodatabase

Next, you'll import LandUse_Pgh.tif into the Chapter10.gdb file geodatabase.

1. **Search for and open the** Raster To Other Format **tool.**

2. **For the Input Rasters field, browse to Chapter10\Data, select LandUse_Pgh.tif, and click OK.**

3. **For Output workspace, browse to Chapter10\Tutorials, select Chapter10.gdb, and click OK.**

4. **For Raster format, select Esri Grid, and run the tool.**

5. **Remove LandUse_Pgh.tif from the Contents pane, add LandUse_Pgh from Chapter10.gdb, and move it below NED in the Contents pane.** Nothing appears to change on the map, but the format of LandUse_Pgh is now Esri Grid, and it is located inside a geodatabase.

Set the geoprocessing environment for raster analysis

Environmental settings affect how geoprocessing is carried out by tools in the current project. You'll set the cell size of raster datasets you create to 50 feet, and you'll use Pittsburgh's boundary as the default mask.

1. **On the Analysis tab, click Environments.**

2. **In the Environments settings for the Raster Analysis category, type** 50 **for Cell Size, and select Pittsburgh for the default mask.**

3.   **Click OK.**

Extract land use using a mask

Next, you'll use the Extract By Mask tool to extract LandUse_Pittsburgh from LandUse_Pgh. The resulting raster dataset will have the same extent as Pittsburgh and therefore be a much smaller file to store than the original.

1.   **Search for and open the** Extract By Mask **tool.**

2.   **Type or make the following selections as shown.** Note that the Extract By Mask tool has you specify the mask layer explicitly, in case you want to override the default mask you set in Environments.
     - **Input raster:** LandUse_Pgh
     - **Input raster or feature mask data:** Pittsburgh
     - **Output raster:** LandUse_Pittsburgh

3.   **Run the tool. ArcGIS Pro applies an arbitrary color scheme to the new layer.**

4.   **Remove LandUse_Pgh from the Contents pane.**

5.   **Open the Catalog pane, expand Databases and Chapter10.gdb, right-click LandUse_Pgh, and click Delete > Yes.**

6.   **Use the Pittsburgh bookmark.**

YOUR TURN

Extract NED_Pittsburgh from NED using the Extract By Mask tool. In the Geoprocessing pane, for Extract By Mask, click Environments and the Select coordinate system button 🌐 on the right of Output Coordinate System. Expand Projected Coordinate System > State Plane > NAD 1983 (US Feet) > NAD 1983 StatePlane Pennsylvania South FIPS 3702 (US Feet) and click OK. Then on the Parameters tab, select NED as input raster, Pittsburgh as the mask, and **NED_Pittsburgh** as the output raster, and run the tool. After creating **NED_Pittsburgh**, remove NED from the map, and delete it from Chapter10.gdb.

Symbolize a raster dataset using a layer file

1. In Contents, move LandUse_Pittsburgh above NED_Pittsburgh, open LandUse_Pittsburgh's Symbology pane, and click the Options button > Import from layer file.

2. Browse to Chapter10\Data, and double-click LandUse.lyr.

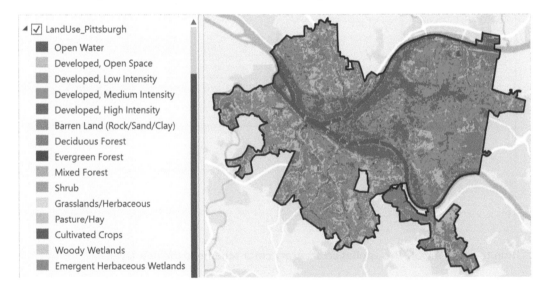

Create hillshade for elevation

Hillshade provides a way to visualize elevation. The Hillshade tool simulates illumination of the earth's elevation surface (the NED raster layer) using a hypothetical light source representing the sun. Two parameters of this function are the altitude (vertical angle) of the light source above the surface's horizon in degrees and its azimuth (east–west angular direction) relative to true north. The effect of hillshade to elevation is striking because of light and shadow. You can enhance the display of another raster layer, such as land use, by making land use partially transparent and placing hillshade beneath it. You'll use the default values of the Hillshade tool for azimuth and altitude. The sun for your map will be in the west (315 degrees) at an elevation of 45 degrees above the northern horizon.

1. **Search for and open the** Hillshade (Spatial Analyst Tools) **tool.**

2. **Type or make the selections as shown.**
    - **Input raster:** NED_Pittsburgh
    - **Output raster:** Hillshade_Pittsburgh

3. **Run the tool.**

Symbolize hillshade

You can improve the default symbolization of hillshade, as you'll do next.

1. **Open the Symbology pane for Hillshade_Pittsburgh.**

2. **Type or make the selections as shown. Click the Color Scheme arrow, click the Show All check box, and select the black to white color scheme.**
    - **Primary Symbology: Classify**
    - **Method: Standard Deviation**
    - **Interval Size: 1/4 standard deviation**

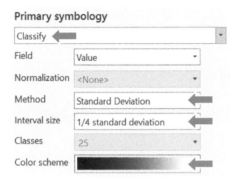

3. **Close the Symbology pane.** The Hillshade looks better, although you can see the 30-meter pixels of the layer.

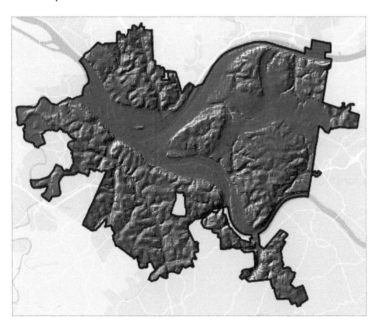

Use hillshade for shaded relief of land use

Next, you'll make LandUse_Pittsburgh partially transparent and place Hillshade_Pittsburgh beneath it to give land use shaded relief.

1. Move **LandUse_Pittsburgh above Hillshade_Pittsburgh, and turn off NED_Pittsburgh.**

2. **Select the LandUse_Pittsburgh layer, and click the Raster Layer contextual tab > Appearance.**

3. **In the Effects group, move the Layer Transparency slider to 33.0 percent.** Hillshade_Pittsburgh shows through the partially transparent LandUse_Pittsburgh, giving the land-use layer a rich, 3D-like appearance.

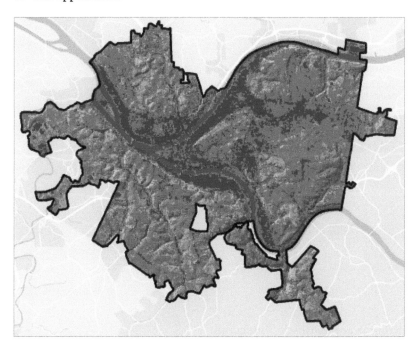

Create elevation contours

Another way to visualize elevation data—elevation contours—is with lines of constant elevation, as commonly seen on topographic maps. For Pittsburgh, the minimum elevation is 215.2 feet and the maximum is 414.4 feet, with about a 200-foot difference. If you specify 20-foot contours, starting at 220 feet, there will be about 10 contours. Note that the output contours are vector line data (and not polygon data).

1. **Search for and open the** Contour (Spatial Analyst) **tool.**

2. **Type or make your selections as shown.** You can use the z-factor to change units. For example, if the vertical units were meters, you would enter 3.2808 to convert to feet. We'll leave the value at 1, because the vertical dimensions are already in the desired units of feet.

    - **Input raster:** NED_Pittsburgh
    - **Output feature class:** Contour_Pittsburgh
    - **Contour interval:** 20
    - **Base contour:** 220

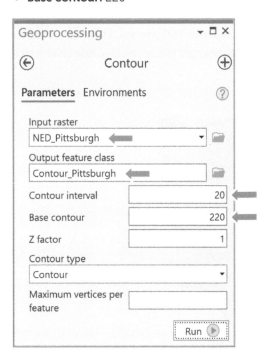

3. **Run the tool.**

4. **Turn off LandUse_Pittsburgh and Hillshade_Pittsburgh.**

5. **Open the Symbology pane for Contour_Pittsburgh, and use Single Symbols for Symbology with a medium-gray line of width** 0.5. Note that you can label contours with their contour elevation values when zoomed in, but you will not do that in this exercise.

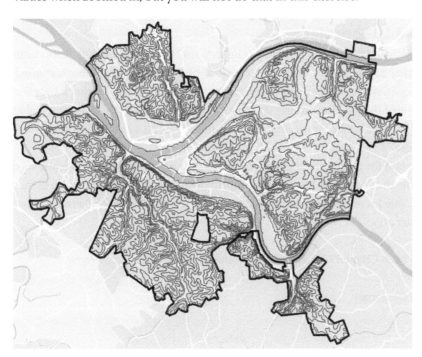

6. **Save your project.**

Tutorial 10-2: Make a kernel density (heat) map

Kernel density smoothing (KDS) is a widely used method in statistics for smoothing data spatially. The input is a vector point layer, often centroids of polygons for population data or point locations of individual demands for goods or services. KDS distributes the attribute of interest of each point continuously and spatially, turning it into a density (or heat) map. For population, the density is, for example, persons per square mile.

KDS accomplishes smoothing by placing a kernel, a bell-shaped surface with surface area 1, over each point. If there is population, N, at a point, the kernel is multiplied by N so that its total area is N. Then all kernels are summed to produce a smoothed surface, a raster dataset.

The key parameter of KDS is its search radius, which corresponds to the radius of the kernel's footprint. If the search radius is small, you get highly peaked mountains for density. If you choose a large search radius, you get gentle, rolling hills. If the chosen search radius is too small (for example, smaller than the radius of a circle that fits inside most polygons that generate the points), you will get a small bump for each polygon, which does not amount to a smoothed surface.

Unfortunately, there are no perfect guidelines on how to choose a search radius, but sometimes you can use a behavioral theory or craft your own guideline for a case at hand. For example, crime hot spots (areas of high crime concentrations) often run the length of the main street through a commercial corridor and extend one block on either side. In that case, use a search radius of one city block's length.

Open the Tutorial 10-2 project

The map in this tutorial has the number of myocardial infarctions (heart attacks) outside of hospitals (OHCA) during a five-year period by city block centroid. One of the authors of this book studied this data to identify public locations for defibrillators, devices that deliver an electrical shock to revive heart attack victims. One of the location criteria was that the devices be in or near commercial areas. Therefore, the commercial area buffers are commercially zoned areas, plus about two blocks (600 feet) of surrounding areas.

KDS is an ideal method for estimating the demand surface for a service or good because its data smoothing represents the uncertainty in locations for future demand, relative to historical demand. Also, in this case, heart attacks, of course, do not occur at block centroids, so KDS distributes heart attack data across a wider area.

1. **Open Tutorial10-2 from Chapter10\Tutorials, and save it as** Tutorial10-2YourName.

2. **Use the Pittsburgh bookmark.**

Run KDS

Blocks in Pittsburgh average about 300 feet per side in length. Suppose that health care analysts estimate that a defibrillator with public access can be identified for residents and retrieved for use as far away as 2 1/2 blocks from the location of a heart attack victim. They recommend looking at areas that are five blocks by five blocks in size (total of 25 blocks, or 0.08 square miles), 1,500 feet on a side, with defibrillators located in the center. With this estimate in mind, you'll use a 1,500-foot search radius to include data within reach of a defibrillator, plus data beyond reach to strengthen estimates.

The objective is to determine whether Pittsburgh has areas that are roughly 25 blocks in area and, as specified by policy makers, have an average of about five or more heart attacks per year outside of hospitals.

1. **Search for and open the** Kernel Density **tool.**

2. **Click Environments, and for Cell size, type** 50, **and for Mask, select Pittsburgh.** Note that environment settings made in the tool apply only to running the tool itself and not to other tools, such as when you set Environments on the Analysis tab.

3. **Click Parameters, and type or make the selections as shown.**
    - **Input point or polyline features:** OHCA
    - **Population field:** TOTAL
    - **Output raster:** HeartAttackDensity
    - **Search radius:** 1500

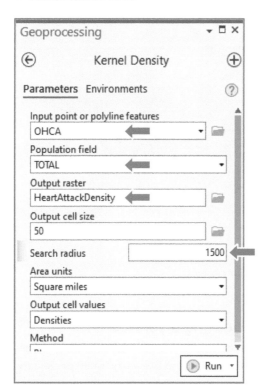

4. **Run the tool.**

5. **Turn off the OHCA and Commercial Area Buffer layers to see the smoothed layer.** The default symbolization is not effective in this case, so next you'll resymbolize the layer.

6. **Symbolize the layer, choose Classify, using Standard Deviation for method with 1/4 standard deviation interval size, and the green to yellow to red color scheme named Condition Number.**

7. **Close the Symbology pane.** Your break points in the legend for HeartAttackDensity may vary somewhat from those in the figure. The smoothed surface provides a good visualization of the OHCA data, whereas the original OHCA points, even with graduated size point symbols for number of heart attacks, are difficult to interpret. Try turning the OHCA layer on and off to see the correspondence between the raw-versus-smoothed data. Note that the densities are in heart attacks per square mile and that the maximum is nearly 1,000 per square mile over five years. A square mile is a large area in a city. The target area of 25 blocks is only 0.08 square miles. Also, only a small part of Pittsburgh (and much less than a square mile) has a density of nearly 1,000 heart attacks per square mile.

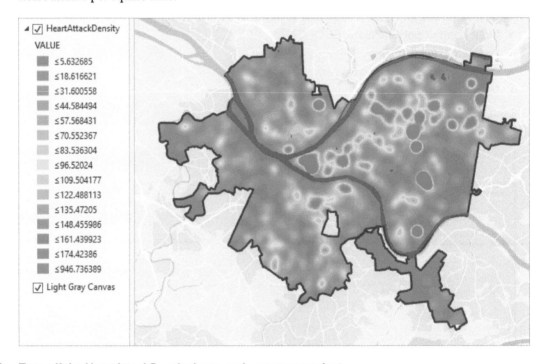

8. **Turn off the HeartAttackDensity layer, and save your project.**

Create a threshold contour layer for locating a service

Assuming that target areas will be around a tenth of a square mile in area, suppose policy makers decide on a threshold of 250 or more heart attacks per square mile in five years (or 50 per square mile per year). Next, you'll create vector contours from the smoothed surface to represent this policy.

1. **Search for and open the** Contour List (Spatial Analyst) **tool.**

2. **Make the following selections or type as shown.** You'll create only one contour, 250, for the threshold defining a target area.
   - **Input raster:** HeartAttackPittsburgh
   - **Output polyline features:** Threshold
   - **Contour values:** 250

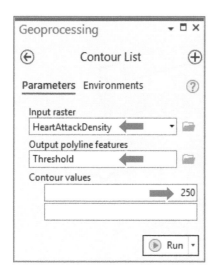

3. **Run the tool.**

4. **Symbolize Threshold with a** 2 **pt bright-red color, turn off Rivers, and turn on Commercial Area Buffer.** Relatively few areas—seven—meet the criterion, and three of those areas appear quite small. All seven threshold areas are in or overlap commercial areas, so you can consider all seven as potential sites for defibrillators. Do any of the areas have the expected number of heart attacks per year to warrant defibrillators? The next exercise addresses this question. Turn the Commercial Area Buffer off.

Estimate the number of annual heart attacks using threshold areas

To estimate annual heart attacks, you can select OHCA centroids within each threshold area, sum the corresponding number of heart attacks, and divide by 5 since OHCA is a five-year sample of actual heart attacks. You will use the Threshold boundaries in a selection by location query, in which case the Threshold layer must be polygons. However, if you examine the properties of the Threshold layer, you'll see that it has the line vector type and not polygon type, even though all seven areas look like polygons. The tool you ran to create Thresholds creates lines because some peak areas of an input raster can overlap with the border of the mask—Pittsburgh, in this case. The lines for such cases would not be closed but left open at the border. Nevertheless, ArcGIS Pro has a tool to create polygons from lines that you'll use next.

1. **Search for and open the** Feature To Polygon **tool.**

2. **Type or make the selections as shown.**
    - **Input Features:** Threshold
    - **Output Feature Class:** ThresholdAreas

    Leave Preserve attributes checked.

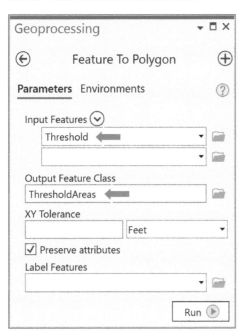

3. **Run the tool.**

4. **Search for the Summarize Within tool (Analysis Tools), and make the selections as shown.** In effect, this tool runs the Summary tool several times—in this case, once for each polygon in the ThresholdAreas feature class to calculate corresponding statistics in the OHCA feature.
    - **Input Polygons:** ThresholdAreas
    - **Input Summary Features:** OHCA
    - **Field:** TOTAL
    - **Statistic: Sum**

Leave Keep all input polygons unchecked.

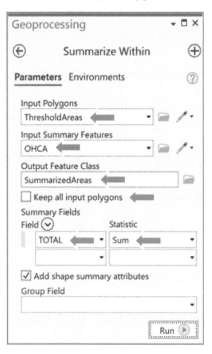

5.   **Run the tool.**

6.   **Open the SummarizedAreas attribute table, and sort by Sum TOTAL descending.** One of the
     ThresholdAreas polygons had no OHCA points inside the polygon but was surrounded by points
     that contributed to its peak density. Clearing the Keep Polygons With No Points check box
     eliminated that polygon from being summarized. The best candidate for a defibrillator has an
     annual average of 104/5 = 20.8 heart attacks per year and is not in the central business district
     (CBD), the area where the westernmost threshold area lies, but is the large area on the east side
     of Pittsburgh. The CBD is in the second row with an average of 68/5 = 13.6 heart attacks per year.
     The last row's area is the only one that does not meet the criterion of at least five heart attacks per
     year on average, with 4.2.

OBJECTID	Shape	Shape_Length	Shape_Area	Count of Points	Sum TOTAL ▾
2	Polygon	7814.828834	4603693.015225	11	104
3	Polygon	7441.571393	3748707.201528	26	68
5	Polygon	5486.377959	2384218.447327	9	45
1	Polygon	3290.86431	860652.955405	2	27
4	Polygon	3579.357673	1018573.404563	2	26
7	Polygon	2183.539239	378075.308945	2	21

The ThresholdAreas polygon that has no OHCA points was not included in the previous table.
The polygon (gray fill color) is about a city block in size and is predicted to be in a peak density
area on the basis of contributions from the kernels of nearby OHCA points, but it has no points
itself. Given the polygon's small size, its 28 nearby heart attacks in the five-year sample, and its

distance within five blocks of the peak area, perhaps the polygon also warrants consideration as a defibrillator site.

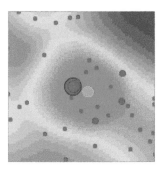

7. **Save your project.**

Tutorial 10-3: Build a risk index model

In this tutorial, you learn more about creating and processing raster map layers. You also are introduced to ArcGIS Pro's ModelBuilder to build models. A model, also known as a macro, is a computer script that you create without writing computer code (a script runs a series of tools). Instead of writing computer code, you drag tools to the model's canvas (editor interface) and connect them in a workflow. Then ArcGIS Pro writes the script in the Python scripting language. Ultimately, you can run your model just as you would any other tool.

The model that you will build in this tutorial calculates an index for identifying poverty areas of a city by combining raster maps for the following poverty indicators:

*   Population below the poverty income line
*   Female-headed households with children
*   Population 25 or older with less than a high school education
*   Number of workforce males 16 or older who are unemployed

Low income alone is not enough to identify poverty areas, because some low-income persons have supplemental funds or services from government programs or relatives, and so rise above the poverty level. Female-headed households with children are among the poorest of the poor, so these populations must be represented when you consider poverty areas. Likewise, populations with low educational attainment and/or low employment levels can help identify poverty areas.

Dawes (1979) provides a simple method for combining such indicator measures into an overall index. If you have a reasonable theory that several variables are predictive of a dependent variable of interest (whether the dependent variable is observable or not), Dawes contends that you can proceed by removing scale from each input and average the scaled inputs to create a predictive index. A good way to remove scale from a variable is to calculate z-scores, subtracting the mean and dividing by the standard deviation of each variable. Each standardized variable has a mean of zero and standard deviation of one (and therefore no scale).

Table 10-1 for Pittsburgh block groups shows that if you averaged the four variables, the variable female-headed households would have a small weight, given its mean of only 36.1, whereas the means of the other three variables are all higher than 100. Z-scores level the playing field so that all variables have an equal role.

Table 10-1

Indicator variable	Mean	Standard deviation
Female-headed households with children population	36.1	41.6
Less than high school education population	116.2	95.2
Male unemployed population	155.3	173.0
Poverty income	167.2	187.0

The following workflow to create the poverty index has three steps:

1. Calculate z-scores for each of the four indicators.
2. Create kernel density maps for all four z-score variables. Although you need the kernel density maps as input to the third step, it is recommended that you also study them individually to understand their spatial patterns in relationship to the poverty index.
3. Use a tool to average the four weighted raster surfaces and create the index raster layer.

Experts and stakeholders in the policy area using the raster index can judgmentally give more weight to some variables than others if they choose. They must use only nonnegative weights that sum to 100. If judgmental weights for four input variables are 70, 10, 10, and 10, the first variable is seven times more important than each of the other three variables. With different stakeholders possibly having different preferences, having a macro allows you to repeatedly run the macro with different sets of weights for the multiple-step process for creating an index raster layer. For example, some policy makers (educators and grant-making foundations) may want to emphasize unemployment or education and give those inputs more weight than others, whereas other stakeholders (human services professionals) may want to heavily weight female-headed households.

You need to standardize the input variables only once, so you'll do that step manually, but you'll complete parts 2 and 3 of the workflow using a ModelBuilder model for creating indexes.

Open the Tutorial 10-3 project

All input variables must come from the same point layer (in this case, block group centroids) as input to build the index so that the data standardization and averaging process is valid. You'll use a 3,000-foot buffer of the study region (Pittsburgh) for two purposes.

First, KDS uses the northernmost, easternmost, southernmost, and westernmost points of its input point layer to define its extent. If the inputs are polygon centroids in a study region, the corresponding KDS raster map will be cut off and not quite cover the study region. The block group centroids added by the buffer yield KDS rasters that extend a bit beyond Pittsburgh's border, but the Pittsburgh mask will show only the portion within Pittsburgh.

Second, in applying KDS, the buffer eliminates the boundary problem for estimation caused by abruptly ending data at the city's edge. KDS estimates benefit from the additional data provided by the buffer beyond the city's edge.

1. **Open Tutorial10-3 from EsriPress\GIST1Pro\Exercises\Chapter10, and save it as** Tutorial10-3YourName.

2. **Use the Pittsburgh bookmark.**

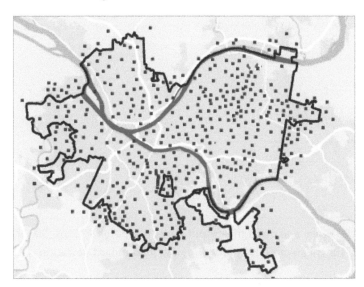

Standardize an input attribute

PittsburghBlkGrps already has three out of four input attributes standardized and ready to use in the poverty index you'll compute, but FHHChld has not yet been standardized. For practice purposes, you will standardize this attribute next.

1. **Open the attribute table of PittsburghBlkGrps, and scroll to the right to see FHHChld (female-headed household with children).**

2. **Right-click the FHHChld column header, click Summarize, and set Statistic Type to Mean and Standard Deviation. Leave Case field blank. Click on the left of the Case field, in the margin, and click its red X to clear it.**

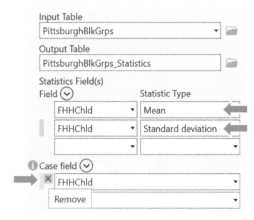

3. **Run the tool. Close the attribute table, and open the output table,** PittsburghBlkGrps_Statistics. Rounded to one decimal place, the mean of FHHChld is 36.1, and its standard deviation is 41.6.

4. **In the Contents pane, right-click PittsburghBlkGrps, and click Design > Fields.**

5. **Scroll to the bottom of the Fields table, and click to add a new field.**

6. **Type** ZFHHChld **for Field Name, select Float for Data Type, click Save, and close the Fields view.**

7. **In the PittsburghBlkGrps attribute table, right-click the ZFHHChld column heading, click Calculate Field, and below the Fields list, create the expression** (!FHHChld! – 36.1)/41.6, **and click Run.** The ZFHHChld value for the record for ObjectID = 1 is −0.146635.

8. **Close the tables, and save your project.**

Set the geoprocessing environment for raster analysis

You'll set the cell size of rasters you create to 50 feet, and you'll use Pittsburgh's boundary as a mask.

1. **On the Analysis tab, click Environments.**

2. **In the Raster Analysis section of Environments, type** 50 **for Cell Size, and select Pittsburgh for the mask.**

3. **Click OK.**

Create a new toolbox and model

A toolbox is necessary to contain models (macros). When your project was created, ArcGIS Pro built a toolbox, named Chapter10.tbx, which is where your model will be saved.

1. On the Analysis tab, in the Geoprocessing group, click ModelBuilder.

2. Open the Catalog pane, expand Toolboxes, expand Chapter10.tbx, right-click Model, and click **Properties.**

3. For Name, type PovertyIndex **(with no space between the two words); for Label, type** Poverty Index; **and click OK.**

4. Hide the Catalog pane.

5. On the ModelBuilder tab, in the Model group, click Save.

Add processes to the model

ModelBuilder has a drag-and-drop environment: you'll search for tools, and when you find them, you'll drag them to your model, open their input/output/parameter forms, and fill out the forms.

1. Search for, but do not open, the Kernel Density tool.

2. Drag the Kernel Density tool to your model.

3. Drag the tool and its output to the upper center of the model window, and click anywhere in the model's white space to deselect it. You'll need a total of four kernel density processes, one for each of the four poverty indicators.

4. **Repeat step 2 three more times, and arrange your model elements as shown in the figure by dragging rectangles around elements to select them and dragging the selections in place.**

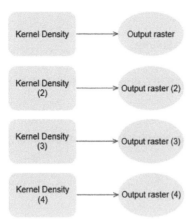

5. **Search for the Raster Calculator tool (Spatial Analyst Tools), drag it to the right of your other model components, and close the Geoprocessing pane.**

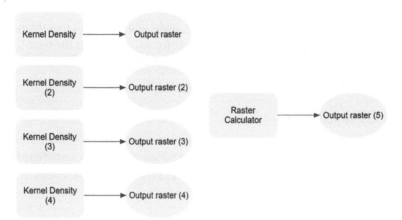

Configure a kernel density process

1. **Double-click the first Kernel Density process, type or make the selections as shown, and click OK.** Fully configured processes and their valid inputs and outputs are filled with color. The 3,000-foot search radius is a judgmental estimate large enough to produce good KDS surfaces at the neighborhood level.
    - **Input point or polyline features:** PittsburghBlkGrps
    - **Population Field:** ZFHHChld
    - **Output raster:** FHHChldDensity
    - **Search radius:** 3000

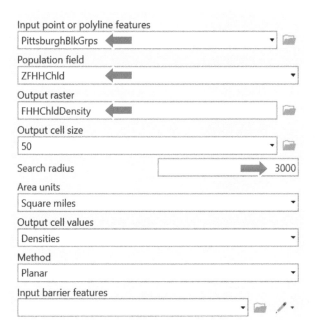

2. **Right-click Kernel Density, and rename it** FHH Kernel Density.

3. **Save your model.**

YOUR TURN

Configure three remaining Kernel Density processes, each with PittsburghBlkGrps as the input, a cell size of **50**, and a search radius of **3,000**, and with population fields **ZNoHighSch**, **ZMaleUnem**, and **ZPoverty**. Refer to the figure for output names. Resize model elements to improve readability. The numbers of the input block groups do not have to match those in the figure. ModelBuilder just needs the names of all model elements to be unique. Rename and resize model objects to make them readable and well aligned. Review all four KDS processes to make sure that they have correct z-score variable inputs and a 3,000-foot search radius. Save your model.

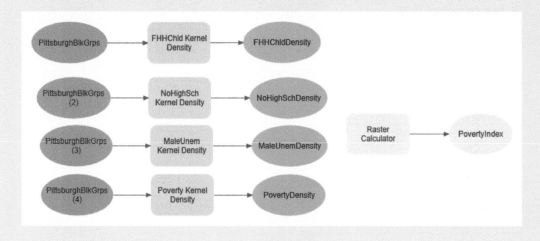

Configure the raster calculator process

1. **Double-click Raster Calculator, and type or make the selections as shown (double-click density rasters to enter them into the expression).** The weights serve as default values for running the model unless the user changes them as parameters for each weight that you will add later in this tutorial.

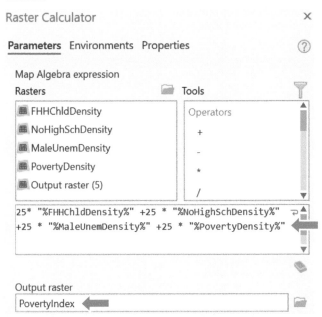

2. **Click OK.** Now the output raster layers from the four kernel density processes are inputs to the Raster Calculator, and all model elements have fill and are ready for running.

3. **Save the model.**

Run a model in edit mode

You can run the entire model by clicking its Run button, which you will do next. When you build a model, however, you can run each process one at a time, following the workflow by right-clicking processes and clicking Run, which allows you to isolate errors and fix them. When you finish, in addition to running the model in Edit mode, you'll also run the model in the Geoprocessing pane, the same way you'd run any tool. ModelBuilder automatically builds a user interface for your model and allows you to change any model elements that you set as parameters.

1. **On the ModelBuilder tab, in the Run group, click Run.** The model runs and produces a log as shown next. If there is an error, the log gives you information about it. Running in Edit mode does not add PovertyIndex to the map. You'll have to add it manually. However, when you run the model as a tool in the Geoprocessing pane, PovertyIndex is added to the map automatically

(when you make PovertyIndex a parameter). If errors arise, fix them, and on the ModelBuilder tab, in the Run group, click Validate, and run the model again.

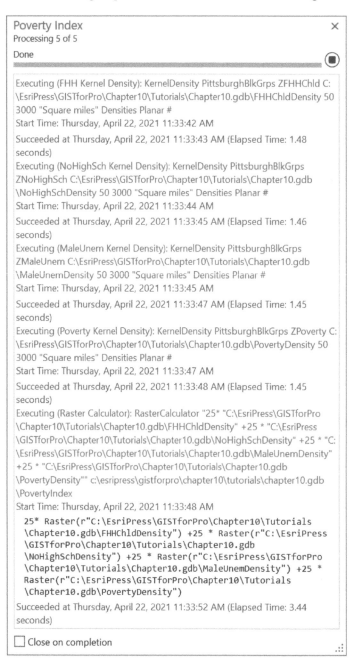

2.   **Close the log file.**

3.   **In the model, right-click PovertyIndex, and click Add To Display.**

4. **Open your map, and turn off PittsburghBlkGrps and Rivers.** You'll improve symbolization of
    PovertyIndex next.

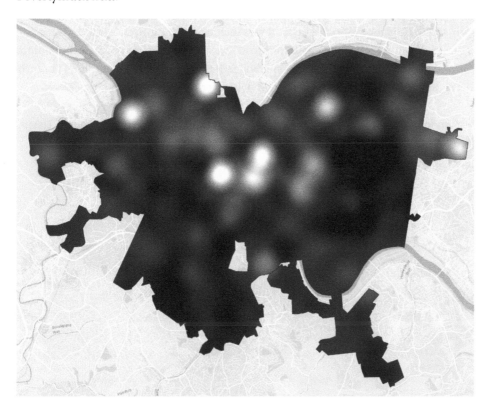

Symbolize a KDS raster layer, and save its layer file

Although it is recommended that vector layers have a maximum of about seven to nine categories for symbolization of numeric attributes (to avoid a cluttered map and allow easy interpretation using a legend), it is advised that symbolization of raster layers include many more categories to represent continuous surfaces. You'll use standard deviations to create categories and the maximum number of categories with that method. Finally, you'll save symbolization as a layer file for automatic use whenever the model creates its output, PovertyIndex, with whatever weights the user chooses.

1. **Open the Symbology pane for PovertyIndex, select Classify for Symbology, Standard Deviation for Method, 1/4 standard deviation for Interval size, and the green to yellow to red color scheme.**

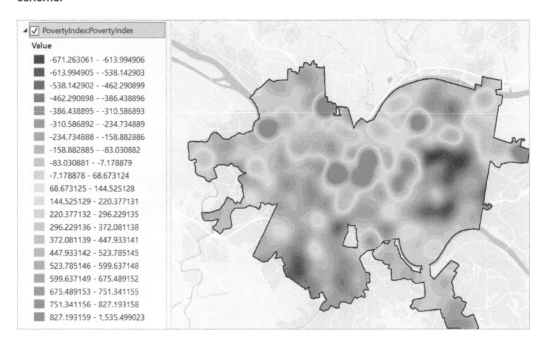

Your break values may be somewhat different from those in the figure.

2. **With Poverty Index's Symbology window still open, change the upper value of the largest category from 1535.499023 to** 3000. The change in upper value is a precaution so that if the weights used to run the model lead to a maximum density greater than 1535, the higher-density pixels will fall in this category.

3. **Close the Symbology pane.**

4. **Right-click PovertyIndex, click Sharing > Save As Layer File, and save as** PovertyIndex.lyrx **to Chapter10\Tutorials.** You'll use the layer file to symbolize PovertyIndex automatically in future runs of the model.

5. **Remove PovertyIndex from the Contents pane.**

Add variables to the model

One objective for the PovertyIndex model is to allow users to change the poverty index's weights. To accomplish this change, you must create variables that will store the weights that the user inputs and designate each variable as a parameter input by the user. ModelBuilder automatically creates a user interface for your model (just like the interface for any tool). The user can enter weights for your model in the interface as an alternative to the default equal weights of 25 each for the poverty index model.

1. **Using Catalog, edit your model.** Notice that each process and output has a drop shadow, indicating that it is finished running.

2. **On the ModelBuilder tab, in the Run group, click Validate.** Validate places all model components in the state of readiness to edit or run again. The drop shadows disappear.

3. **On the ModelBuilder tab, in the Insert group, click the Variable button, select Variant for Variable Data Type, click OK, and move the Variant variable above the Raster Calculator process.** ArcGIS Pro can determine the actual data type of a variant data type variable from entered values. For example, when you give the variable the value 25 in step 4, ArcGIS Pro will treat the value as an integer variable.

4. **Right-click the variable, click Open, type** 25, **and click OK.** The value of 25 is the default value for running the model if the user does not change it.

5. **Right-click the Variant variable, click Rename, type** FHHChldWeight, **and press Enter.**

6. **Right-click FHHChldWeight, and click Parameter.** That action places a letter *P* near the variable, making the variable a parameter. In other models you build, consider making the model inputs (in this case, the four rasters) parameters if they can change from run to run. Then users can browse for these parameters when they run the model as a tool.

YOUR TURN

Add three more variant variables, all with value **25**, and named **NoHighSchWeight**, **MaleUnemWeight**, and **PovertyWeight**. Make each variable a parameter.

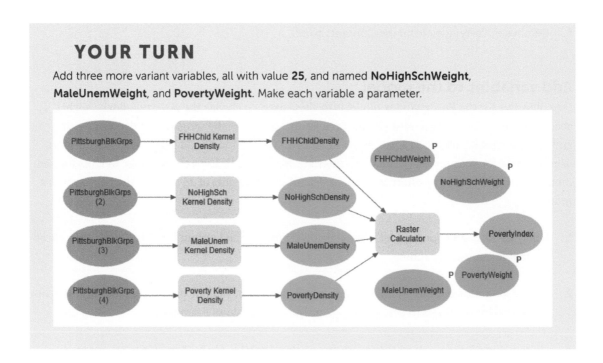

Use in-line variable substitution

In this exercise, you will transfer the weights stored in variables (values that will be input by the user) to parameters of the Raster Calculator tool. The mechanism is called in-line substitution because the variables' values are substituted for the model's parameter values.

1. Open the Raster Calculator process in the model.

2. Select and delete the first 25 in the expression, leave the pointer in its current position at the beginning of the expression, scroll down in the Rasters pane, and double-click FHHChldWeight to enter it into the expression where the 25 had been.

3. Delete the double quotation marks around "%FHHChldWeight%".

4. Likewise, select and replace the 25 weights for the remaining rasters, yielding the finished expression as follows:

    ```
    %FHHChldWeight% * "%FHHChldDensity%" + %NoHighSchWeight% *
    "%NoHighSchDensity%" + %MaleUnemWeight% * "%MaleUnemDensity%" +
    %PovertyWeight% * "%PovertyDensity%"
    ```

5. Click OK. Your Raster Calculator process and its output will have color fill.

6. On the ModelBuilder tab, in the Run group, click Validate.

7.   Save your model.

Use a layer file to automatically symbolize the raster layer when created

You'll use the layer file you created earlier for the poverty index for this task. To do so, you must make the PovertyIndex output of the Raster Calculator process a parameter in the model.

1.   **Make PovertyIndex a parameter.** Making PovertyIndex a parameter also adds it to the Contents pane for display in your map when you run the model as a tool from the Geoprocessing pane.

2.   **Open the Catalog pane, expand Toolboxes > Chapter10.tbx, right-click Poverty Index, and click Properties > Parameters.** Note that frequently you must click Refresh when viewing geodatabases to see the latest additions.

3.   **Scroll to the right, and under the Symbology column heading, click the cell for the row with the label PovertyIndex.**

4.   **In that cell, click the resulting Browse button, browse to Chapter10\Tutorials, and double-click PovertyIndex.lyrx.**

	Label	Name	Data Type	Type	Direction	Cat	Filt	Dep	Default	Env	Symbology
0	FHHChldWeight	FHHChldWeight	Variant	Required	Input				25		
1	NoHighSchWeight	NoHighSchWeight	Variant	Required	Input				25		
2	MaleUnemWeight	MaleUnemWeight	Variant	Required	Input				25		
3	PovertyWeight	PovertyWeight	Variant	Required	Input				25		
4	PovertyIndex	PovertyIndex	Raster Dataset	Required	Output				C:\EsriPres...		C:\EsriPress\GISTforPro\Chapter10\Tutorials\PovertyIndex.lyrx

5.   **Click OK, and save and close your model.**

Run your model

Congratulations, your model is ready to use.

1.   **With your map open, ensure that PovertyIndex is removed from the Contents pane.**

2. **On the Analysis tab, click Tools, and search for and open your** Poverty Index **model.**

3. **Leave the weights at their default settings, and run the model.** See that the model computes the poverty index, adds the index to the Contents pane, and symbolizes it with your layer file.

4. **Save your project.**

YOUR TURN

Remove PovertyIndex from the Contents pane, and run your model with different weights (non-negative and sum to 100), such as 70, 10, 10, and 10 saving the output as **PovertyIndex_Chld70**. See how the output changes a lot. Note that for a class project or work for a client, you must symbolize PovertyIndex manually to get the best color scheme and categories for a final model, depending on the distribution of densities produced. When you finish, save your project, and close ArcGIS Pro.

Assignments

This chapter has two assignments to complete that you can download from ArcGIS Online, at go.esri.com/GISTforPro2.8Data:

- **Assignment 10-1:** Create raster maps for the Pittsburgh Almono development area.
- **Assignment 10-2:** Estimate heart attack fatalities outside of hospitals in Wilkinsburg by gender.

References

Dawes, Robyn M. 1979. "The Robust Beauty of Improper Linear Models in Decision-Making." *American Psychologist* 34 (7): 571–82.

3D GIS

LEARNING GOALS

- Explore global scenes.
- Learn how to navigate scenes.
- Create local scenes and TIN surfaces.
- Create z-enabled features.
- Create 3D buildings and bridges from lidar data.
- Work with 3D features.
- Use procedural rules and multipatch models.
- Create an animation.

Introduction

This chapter introduces ArcGIS Pro's 3D display and processing of data and maps. 3D maps provide insights that are not readily apparent from 2D visualization of the same data. For example, instead of inferring the presence of a valley from 2D contours, in 3D you see the valley as well as the difference in height between the valley floor and a ridge. 3D maps allow you to design, visualize, communicate, and analyze for better decision-making.

Navigation of ArcGIS Pro is similar in 2D and 3D; however, there are differences when you explore maps with data, such as raster surfaces, lidar data, and 3D features. ArcGIS Pro uses 3D scenes and one of two viewing modes. The first, global, is for large-extent, real-world content in which the curvature of the earth is important. The second, local, is for smaller-extent content in a projected coordinate system or cases in which the curvature of the earth isn't important. You can switch scenes between global and local. You can also modify a scene's settings, depending on the scale of the project. Rendering 3D scenes is slower than rendering 2D maps, and proper computer hardware and configuration is necessary.

This chapter explains procedural methodologies that allow rapid creation of 3D content. The methodologies describe how to construct multiple 3D models on the basis of feature attributes—such as building heights and roof shape types—rather than creating a single, specific 3D model. Procedural rules, which define patterns, are authored in ArcGIS® CityEngine®, a 3D modeling software for urban environments. You can reuse procedural rules in other parts of the ArcGIS Platform after they have been exported as rule-package files.

This chapter also explains the use of lidar data and visual analysis tools such as line of sight and introduces 3D animation.

Tutorial 11-1: Explore a global scene

Global scenes use a default elevation surface, WorldElevation3D/Terrain3D, from an ArcGIS Online map service, and coordinate system GCS WGS84. In this tutorial, you will explore a global scene's properties and learn how to navigate in 3D. Advantages of the global scene include working in large or multiple geographic areas, enhanced illumination and time effects, and publishing 3D content to a web scene.

Open the Tutorial 11-1 project

1. **Open Tutorial11-1.aprx from Chapter11\Tutorials, and save it as** Tutorial11-1YourName.aprx. The project opens with a 3D scene using a default elevation surface, Terrain3D, from an Esri map service. In the Contents pane, 3D Scene is labeled with a globe icon, indicating that it's a global scene. A global scene whose extent is not clipped includes the entire surface of the earth.

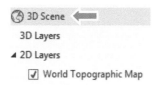

2. **Use the Study Area bookmark.** The map zooms to a study area of the city of Pittsburgh, including the Central Business District, North Shore, South Shore, and Mount Washington neighborhoods. The map includes no added GIS features yet, just a basemap.

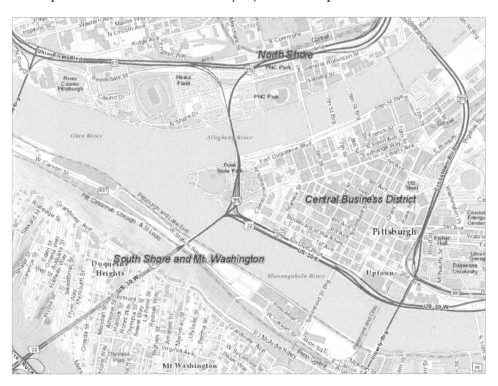

Explore a scene's properties

The scene's elevation surface is different from a basemap, and understanding the elevation surface, map units, and heights is important in a 3D scene.

1. **In the Contents pane, right-click WorldElevation3D/Terrain3D > Properties.**

2. **In the Elevation Source Properties window, click Source.** The scene's elevation source is visible, showing the elevation Service Name, WorldElevation3D/Terrain3D, and its location from ArcGIS.com.

3. **Under Vertical Units, choose Feet, and click OK.** The map and display units will remain decimal degrees for now—only the elevation units change.

Navigate a scene with a mouse and keyboard keys

Next, view the map using a predefined 3D bookmark, and explore using mouse and keyboard shortcuts.

1. **Use the Rivers bookmark.** The view shows that the terrain is higher in the Mount Washington neighborhood, on the left of the view.

Sometimes, you can get disoriented in a 3D view, so learning a few useful shortcuts can help you return to a familiar orientation. Experiment with the following keyboard shortcuts, commonly used to manipulate a 3D view, in the next set of steps.

2. **Press and drag the wheel button to adjust (tilt) the view.**

3. **On the keyboard, press the J or U key to move the map up or down.**

4. **Press the A or D key to rotate the view clockwise or counterclockwise.**

5. **Press the W or S key to tilt the camera up and down.**

6. **Press the left, right, up, or down arrow keys to move the view.**

7. **Press the B key and use the left mouse or arrow keys to look around your view.**

8. **Press the N key to view true north.**

Chapter 11: 3D GIS

GIS Tutorial for ArcGIS Pro 2.8

331

Part

3

Chapter

11

Tutorial

1

9. **Press the P key to look straight down at your map.**

Change the basemap

You can display various basemaps with the current surface elevation. If you want to see imagery details, they will be draped to the elevation surface.

1. **Use the Football Stadium bookmark, and zoom out a few times if necessary to see the Heinz Field football stadium along the river.**

2. **Change the basemap to Imagery.** The imagery drapes to the elevation surface, showing the football stadium, rivers, and trees along the hills above Pittsburgh's South Shore, and so on.

3. **Use the Baseball Stadium bookmark, zoom out, and use the mouse or keyboard to see additional views.**

4. **Change the basemap back to Topographic, and use the Rivers bookmark.**

YOUR TURN

Explore another geographic area, perhaps your hometown, your favorite vacation spot, or a city or an area you have always wanted to visit.

Exaggerate and apply a shade and time to a surface

Sometimes, subtle or important changes in the landscape can be emphasized by adding visual effects to the layer. For example, you can graphically exaggerate the height of a mountainous area to help it stand out. This exaggeration does not actually change the elevation but visually makes features more prominent. Another effect includes adding lighting or illumination sources through shading or by time of day.

1. **In the Contents pane, under Elevation Surfaces, click Ground, and click the Appearance tab.**

2. **For Vertical Exaggeration, type** 3.00, **and turn on Shade Relative to Light Position.**

3. **In the Contents pane, right-click 3D Scene, and click Properties. In the Map Properties dialog box, click Illumination, and for Illumination Defined By, click Date And Time, and click OK.** The elevation will be exaggerated, and the sun shadows are visible and depend on the date and time selected.

YOUR TURN

Pan and navigate the scene to see the exaggeration from different views. Navigate to another area you know is mountainous. Use the Rivers bookmark to return to Pittsburgh. Save your project.

Tutorial 11-2: Create a local scene and TIN surface

Advantages of local scenes include using your own elevation surface data such as TIN (triangulated irregular network) or lidar data, using a projected coordinate system, managing features below a surface (for example, subways or waterlines), and perhaps more accessibility to edit data. You can also set the coordinate system for a local scene to local coordinates (for example, state plane) and use the surface offline. In this tutorial, you will create a TIN surface from contours, change its symbology, and use it as the elevation surface in a local scene.

Open the Tutorial 11-2 project

1. **Open Tutorial11-2.aprx from Chapter11\Tutorials, and save it as** Tutorial11-2YourName.aprx. The project opens with a 3D scene named TIN Surface Scene with 2D layer contours, street curbs, parks, and rivers draped to the default Terrain3D elevation surface in a global scene. The base-map is Light Gray Canvas Base, covering the entire earth. You will convert the global scene to a local scene and clip the basemap to the study area. Unlike the geoprocessing Clip tool, this process clips the layers for display purposes only.

Set a local scene

1. **On the View tab, in the View group, click the Local button.** The scene switches from a global to a local scene, and the icon in the Contents pane and on the view updates. Next, you clip the base layer to the study area using the Contours layer.

2. **In the Contents pane, right-click TIN Surface Scene > Properties > Clip Layers.**

3. **Click Clip To A Custom Extent, and for Get Extent From, click Contours. Click Apply, and click OK.**

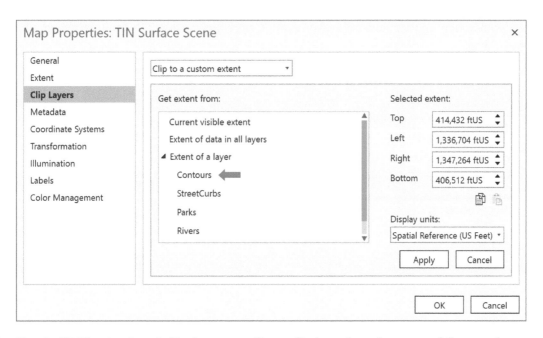

4. **Use the 3D View bookmark.** The basemap will now display only to the extent of Contours layer.

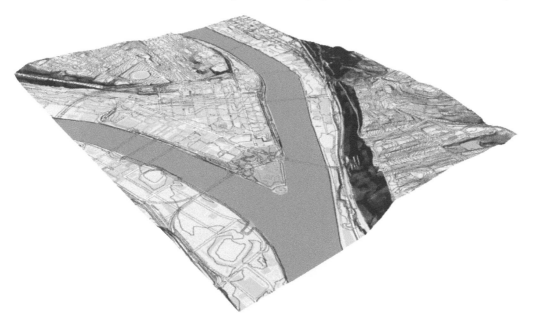

Create a TIN surface

A TIN surface is a vector data model composed of irregularly distributed nodes and lines that are formed from x-, y-, and z-values and arranged in a network of triangles that share edges. TINs are typically used for high-precision modeling of small areas, such as in engineering applications, in which they are useful because they allow calculations of surface area and volume. TIN surfaces are also useful to view underground features or utilities. Here, you create a TIN from the topography contour map.

1. **Search for and open the** Create TIN **tool.**

2. **Type or make the selections as shown.**
    - **Output TIN:** PGH_TIN
    - **Coordinate System:** NAD_1983_StatePlane_Pennsylvania_South
    - **Input Features:** Contours

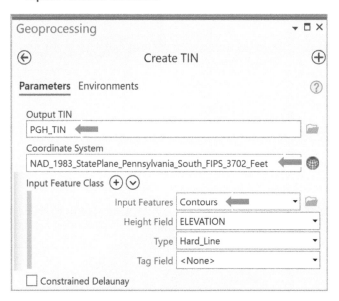

3. **Run the tool.**

4. **In the Contents pane, turn off Contours and the basemap, and expand PGH_TIN.** The TIN elevation is displayed with elevation heights from high to low.

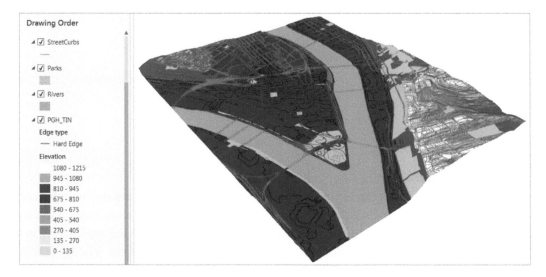

Change the scene's surface and coordinate system

The local TIN surface can now replace the surface assigned from the map service.

1. In the Contents pane, under Elevation Surfaces, right-click Ground > Add Elevation Source, browse to Chapter11\Tutorials, click PGH_TIN, and click OK. The dataset PGH_TIN is the first surface listed as an elevation source.

2. In the Contents pane, under Elevation Surfaces, remove WorldElevation3D/Terrain3D. This step removes this surface as an elevation source.

3. Remove World Light Gray Canvas Base from the Contents pane. Wait for the scene to redraw its features. The scene is now set to local coordinates and TIN as the surface data that could be used offline.

> ### YOUR TURN
>
> Use feet or meters for TIN units, not decimal degrees. Change the scene's properties for display and elevation units to feet, the default unit for NAD 1983 State Plane Pennsylvania South.

Change the TIN's symbology

You can change the symbology of a TIN to better reflect features that the surface model represents. Next, you add the contour and slope symbology renderers.

1. In the Contents pane, click PGH_TIN.

2. On the Appearance tab, in the Drawing group, click the Symbology down arrow and the Slope button ◪.

3. In the Symbology pane, under Draw Using, click Simple. Then use the gallery to change the color of the current symbol color to Land.

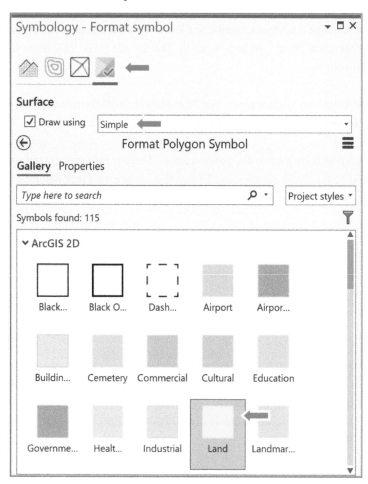

4. Close the Symbology pane, and zoom to better see the features at all elevations.

5. Click the wheel button to view the scene from different angles, including below the scene.

6. Save your project.

Tutorial 11-3: Create z-enabled features

You can create 3D content in different ways, and the corresponding workflows depend on the type of features you create. In addition to creating 3D features from scratch, you can import 3D models and symbolize 2D features as 3D features. You can also specify the source of your z-values when you create features. ArcGIS Pro's Current Z Control is used to set the 3D elevation source for drawing or obtaining z-values. This option is useful if more than one source is defined for a global or local scene, or if you have another source not already included in the map.

In this tutorial, you'll create a 3D feature class that is z-aware. Then you'll use the Current Z Control tool to set the elevation source for populating z-values. The Current Z Control has two modes, constant and surface. Constant is used to create 3D features at an absolute height by typing in an exact value—for example, a plane flying at a constant altitude. Surface uses z-values from the active elevation source you choose.

Open the Tutorial 11-3 project

1. **Open Tutorial11-3.aprx from Chapter11\Tutorials, and save it as** Tutorial11-3YourName.aprx. The map opens with 3D Trees Scene, a local scene using World Topographic as the elevation surface whose extent is clipped to a Rivers layer and that has a Parks layer draped to the surface.

2. **Use the Point State Park bookmark.** The map zooms to a large park at the confluence of Pittsburgh's three rivers.

Create a z-enabled feature class for park trees

To take advantage of certain 3D editing capabilities, you must ensure that an output feature class will have z-values. The Create Feature Class geoprocessing tool allows you to determine these settings. Here, you'll create an empty 3D feature class, digitize new features, and populate z-values directly from the map's surface. If you need to confirm whether a layer is z-enabled, you can verify the data source information listed on the Source page from the Layer Properties dialog box.

1. **Search for and open the** Create Feature Class **tool.**

2.   **Type or make the selections as shown.**
   -   **Feature Class Name:** ParkTrees
   -   **Geometry Type: Point**
   -   **Has Z: Yes**
   -   **Coordinate System: NA_1983_StatePlane_Pennsylvania_South**

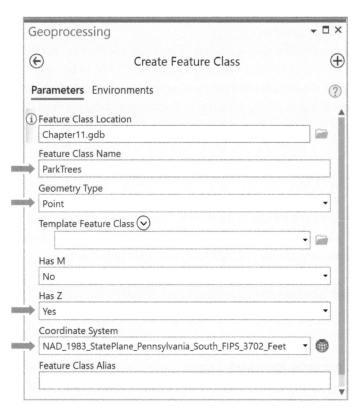

3.   **Run the tool.**

4.   **Change the ParkTrees symbol to Circle 1 and color to Fir Green.**

Digitize trees on surfaces using Z Mode

1.   On the Edit tab, in the Elevation group, click the Z Control button (Z Mode) ⌖z. This step turns on the Z Control option.

2.   **Click the Get Z From View button** ⊥z.

3.   **Click a point inside the center of Point State Park as shown in the figure to set the z-value (height).**

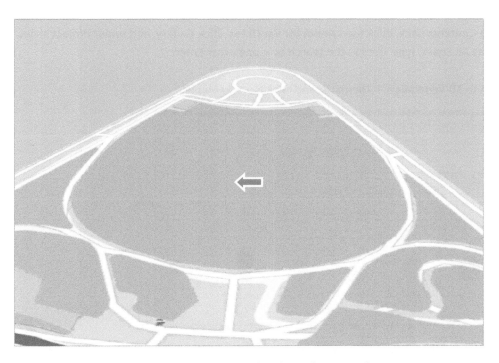

Depending on where you click, the elevation height is about 723 feet.

4. **On the Edit tab, in the Features group, click the Create button.**

5. **In the Create Features pane, select ParkTrees.**

6. **Click about 10 points to digitize trees on each side of the park.** It may take a moment to see each point as you digitize it.

7. **Save your edits, and clear the selection.**

YOUR TURN

Use the Mt. Washington Park bookmark, set the z-value to various elevations along the large park below Mount Washington, and digitize about 20 trees at various elevations along the hill. Save your edits, and turn off the Z Control.

Use 3D symbols with real-world coordinates

You can display 3D tree symbols as low-resolution, high-resolution, or thematic trees. Depending on the number of features and map purpose, you will want to experiment with all three tree types.

1. **Zoom to a few trees on one side of Point State Park.**

2. In the Contents pane, click the symbol for ParkTrees, click Gallery, and under Project styles, choose All Styles, type Trees in the search box, and press Enter.

3. Scroll to 3D Vegetation – Thematic, and click Norway Maple.

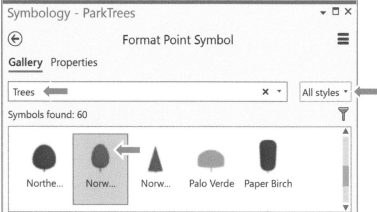

4. **Close the Symbology pane, and zoom in and out.** Trees get bigger and smaller depending on the zoom. Next, you will set a fixed height using real-world units and turn on shading for more realistic trees.

5. In the Contents pane, right-click ParkTrees > Properties > Display, check the box next to Display 3D Symbols In Real-World Units, and click OK.

6. In the Contents pane, right-click 3D Trees Scene, and click Properties. In the Map Properties dialog box, click Illumination, confirm Display Shadows In 3D is checked, and click OK. Next, you'll enter an exact height for the park trees.

7. In the Contents pane, click the Park Trees symbol. In the symbology pane under Properties > Appearance, type 20 (m) for Size, click Apply, and close the Symbology window.

8. **Zoom in and out.** Your scene is now populated with thematic trees at a consistent height.

Add realistic preset trees using a table of a tree's genus (type)

ArcGIS Pro provides preset layers for displaying features in a 3D map. For example, if you have a field with the genus or general type of tree (for example, pine or *Pinus*), you can display multiple trees by their type.

1. On the Map tab, in the Layer group, click Add Preset > Realistic Trees.

2. Browse to Chapter 11 > Tutorials > Chapter11.gdb, select StreetTrees, and click OK.

3. In the Symbology pane, under Type, click GenusName, and close the Symbology pane.

4. **Use the Street Trees bookmark.** Street trees are draped to various elevations and displayed using the tree genus type.

5. Zoom out and pan the map to see more street trees.

6. Save your project.

Tutorial 11-4: Create features and line-of-sight analysis using lidar data

Lidar uses pulsed laser light from aircraft or drones to provide detailed elevation data and classification of land cover that you can use to create 2D surfaces and 3D features. Geographic lidar data is commonly available as lidar aerial survey (LAS) files, the industry standard of the American Society for Photogrammetry and Remote Sensing. In this tutorial, LAS files were provided by Pictometry International Corporation for a study area of Allegheny County, Pennsylvania.

The generation of 3D buildings from lidar LAS datasets requires two surface models, a digital surface model (DSM) and a digital terrain model (DTM), to create a normalized surface (nDSM), which is the difference between the DSM and DTM surfaces used to calculate building heights. The nDSM is applied to random points that are created for 2D building footprints. These footprints are used to generate z-values (heights) for each random point. The z-value of the highest point is the building height. The process finishes by creating a statistics table, which selects the maximum z-value of the random points, that is then joined to 2D building footprints, allowing for buildings to be extruded using that value. You can also use lidar data to determine line-of-sight obstructions between features, such as buildings as seen in the last exercise of this tutorial.

Open the Tutorial 11-4 project with a 3D scene

1. **Open Tutorial11-4.aprx from the Chapter11\Tutorials folder, and save it as** Tutorial11-4YourName .aprx. The project opens with a 3D scene and the building footprints displayed as a 2D layer and a World Light Gray basemap in a local view, with 2D layers for buildings and a lidar delivery area used to clip the basemap. There is no height value in the building attribute table, so you can display buildings only as flat 2D polygons for now. Additional layers of observer points used for line-of-sight analysis and a 3D layer for bridges are turned off.

2. **Use the 3D View bookmark.**

Create a LAS dataset

LAS files have points classified as bare earth, vegetation, buildings, and so on, and are created from original large LAS (American Standard Code for Information Interchange, or ACSII) files, which you can view in 3D or make into raster layers.

A LAS dataset, created from original LAS data, provides fast access to lidar data without the need for data conversion, making it easy to work with LAS files for a specific study area.

1. **Search for and open the** Create LAS Dataset **tool.**

2. **Under Input files, click the Browse button, browse to Chapter 11 > Data > LASFiles, and select the six LAS files.**

3. **Type or make the changes as shown.**
    - **Output LAS Dataset:** Chapter11LasDataset.lasd
    - **Coordinate System: NAD_1983_StatePlane_Pennsylvania_South**
    - **Check Compute Statistics.**

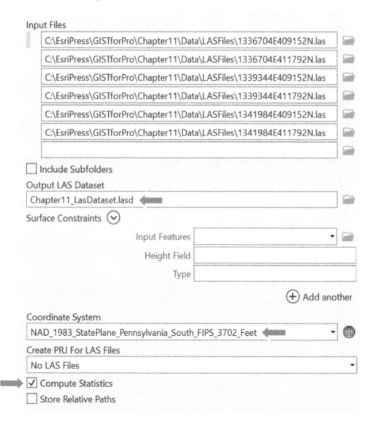

4. **Run the tool.** The lidar data values are clearly shown as points and their values. You can symbolize the LAS dataset using elevations, slope, aspect, and so on. Pittsburgh's tallest building is the US Steel building, the triangular building on the right side of the study area.

5. **Explore the 3D map from various locations, and use the 3D View bookmark.**

Generate a raster DSM (digital surface model)

DSMs represent the surface of the earth, including buildings, tree canopies, and other obstructions. Before generating the DSM raster, you'll first filter lidar points to save processing. You will create a DSM using an interpolation type of binning, which is faster for processing, and a maximum cell assignment to find the highest elevation point within each cell.

1. **In the Contents pane, right-click Chapter11_LasDataset.lasd > LAS Filters, and click 1st Return.**

2. **Search for and open the** LAS Dataset To Raster **tool.**

3. **Type or make selections as shown.**
   - **Input LAS Dataset:** Chapter11_LasDataset.lasd
   - **Output Raster:** DSM
   - **Interpolation Type: Binning**
   - **Cell Assignment: Maximum**
   - **Sampling Value:** 5

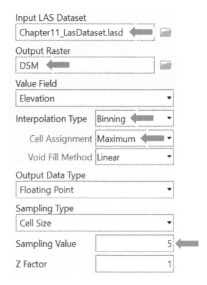

4. **Run the tool.**

5. **Turn off the Chapter11_LasDataset.lasd and Bldgs layers.** The values of the DSM show the range of elevations from high to low.

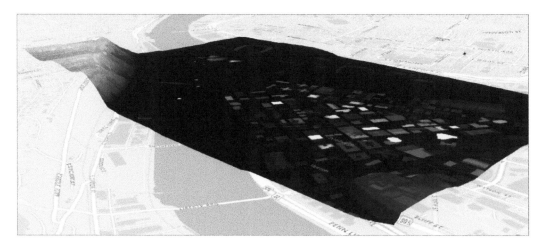

Generate a raster DTM (digital terrain model)

Next, you create the DTM, a bare-earth terrain surface, containing only the topology. In many cases, a DTM is the same as a digital elevation model (DEM). Before creating the raster, you filter the ground features.

1. **Turn off the DSM layer.**

2. **Turn on the Chapter11_LasDataset.lasd layer, and in the Contents pane, right-click Chapter11_ LasDataset.lasd > LAS Filters, and click Ground.** This will filter and show only the ground features used to create the DTM.

3. **Search for and open the** LAS Dataset To Raster **tool, and type or make the selections as shown.** This raster uses a different interpolation type of triangulation that takes a little longer to process but better interpolates earth surface voids found in the LAS Dataset.

   - **Input LAS Dataset:** Chapter11_LasDataset.lasd
   - **Output Raster:** DTM
   - **Interpolation Type: Triangulation**
   - **Interpolation Method: Natural Neighbor**
   - **Sampling Value:** 5

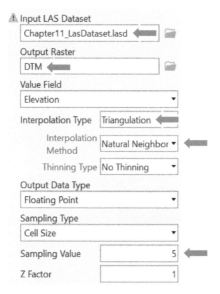

4. **Run the tool and ignore any warning messages.** A raster surface of the earth's features is created.

Create a normalized digital surface model (nDSM) raster

An nDSM surface is the difference between the DSM and DTM surfaces that is normalized to the bare-earth surface.

1. **Search for and open the** Minus (3D Analyst Tools) **tool.**

Chapter 11: 3D GIS

GIS Tutorial for ArcGIS Pro 2.8

349

Part

3

Chapter

11

Tutorial

4

2. **Type or make the selections as shown.**
   - **Input raster or constant value 1:** DSM
   - **Input raster or constant value 2:** DTM
   - **Output raster:** nDSM

Input raster or constant value 1
DSM

Input raster or constant value 2
DTM

Output raster
nDSM

3. **Run the tool. You now have a raster surface that you can apply to point features used for buildings to determine their height.**

Create random points for buildings

Randomly generated points are created for each building polygon, and the nDSM raster surface is applied to each random point. The point with the highest z-value will be used as the building height.

1. **Turn off the Chapter 11_LasDataset layer, and turn on the Bldgs layer.**

2. **Search for and open the** Create Random Points **tool.**

3.   **Type or make the selections as shown.**
   - **Output Point Feature Class:** BldgRandomPoints
   - **Constraining Feature Class:** Bldgs
   - **Number of Points:** 100
   - **Minimum Allowed Distance:** 5

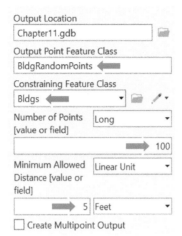

4.   **Run the tool.**

YOUR TURN

Turn off all layers except BldgRandomPoints, and zoom in. Click various random points on each
building. Note that the 100 points for the US Steel building will have a CID value (unique value
for each building) of 521. Every building has a unique CID value that you will use later to join to
building footprints.

Add surface information to random points

Here, you assign z-values (height) from the nDSM raster surface to each random point using the Add
Surface Information tool.

1.   **In the Geoprocessing pane, search for and open the** Add Surface Information **tool.**

2. **Type or make the selections as shown.**
   - **Input Features:** BldgRandomPoints
   - **Input Surface:** nDSM
   - **Output Property: Z checked**

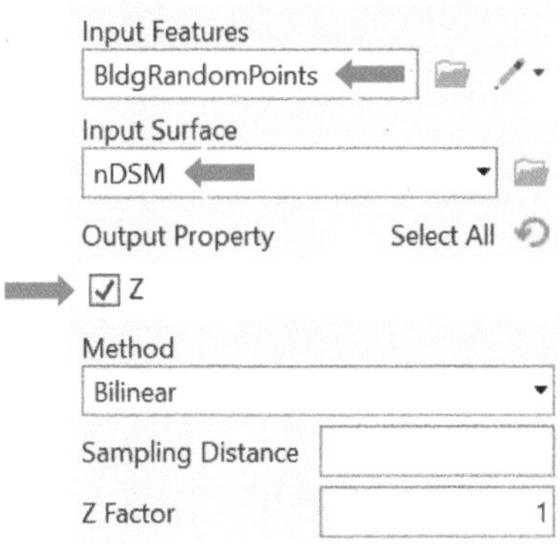

Input Features

BldgRandomPoints ⬅

Input Surface

nDSM ⬅

Output Property   Select All

☑ Z

Method

Bilinear

Sampling Distance

Z Factor   1

3. **Run the tool.**

YOUR TURN

Click on various random points, and note that they now have z-values. The highest value for each building will be used for the building height.

Assign a maximum value (height) to random points

The Summary Statistics tool will calculate the maximum z-value for all buildings using the building's random points. A text file is created that you will join back to buildings using the CID field.

1. **Search for and open the** Summary Statistics **tool.**

2. **Type or make the selections as shown.**
    *   **Input Table:** BldgRandomPoints
    *   **Output Table:** BldgHeights
    *   **Statistics Field:** Z
    *   **Statistic Type: Maximum**
    *   **Case field: CID**

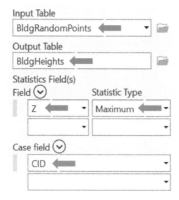

3. **Run the tool.**

YOUR TURN

Open the BldgHeights table, and sort the MAX_Z field in descending order. Look for CID 521; the tallest building will be the US Steel building. Close the table.

Join the maximum z-value (height) to building footprints, and display as 3D buildings

1. **In the Contents pane, right-click Bldgs > Joins and Relates, and click Add Join.**

2. **Join BldgHeights to Bldgs using** BLDG_ID **for Input Join Field and** CID **for Output Join Field.**

3. **Turn on the Bldgs layer, and drag it to 3D Layers in the 3D scene.**

4. **Turn off BldgRandomPoints, and use the 3D View bookmark.**

5. On the Appearance tab, in the Extrusion group, under Type, click Max Height, and select **[MAX_Z] as the field.** These steps required a lot of processing, but you now have 3D buildings. Notice the residential buildings on the left of the view in the Mount Washington neighborhood, as opposed to the taller high-rise buildings in the Central Business District neighborhood.

6. Save your project.

Use lidar to determine bridge elevation heights

You can view and select lidar data points to determine the elevation height to draw bridges. Pittsburgh has more than 750 bridges, but you will use data to find the height and digitize just one.

1. Turn off the Bldgs layer, and use the Fort Pitt Bridge bookmark. Turn on the Chapter11_ LasDataset.lasd layer, and set the LAS Filters to All Points.

2. On the Map tab, click the Explore button, and click various lidar points along the bridge to see the **z-values (height)**. Points at the top of the bridge span are approximately 920 feet, whereas points at the bottom deck range from 770 feet to 800 feet. You will use 775 feet as the elevation to draw the base of the bridge.

Draw a bridge using z-elevation

It's easier to draw the bridge in a 2D map, and you can do so by setting the Z Mode elevation.

1. On the View tab, click Convert > To Map, and turn on the Bridges layer.

2. On the Edit tab, in the Elevation group, click the Z Mode button, and type 775 as the constant elevation.

3. Click Create, and from the Create Features pane, choose Bridges.

4. **Zoom to the bridge, as shown in the figure, and digitize the approximate location of the bridge.**

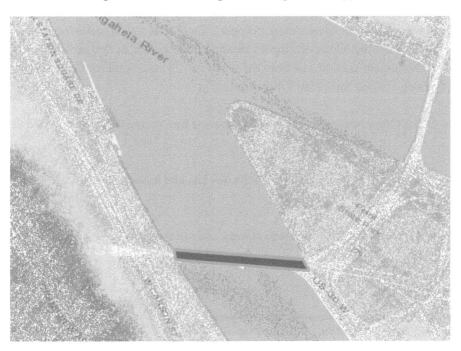

5. **Save your edits, and close the 2D map to return to the 3D scene and turn the Bridges layer on.** The bottom of the bridge will be at the correct elevation. You can also snap lidar points to create 3D features.

6. **Click the Create button, and digitize a bridge span polygon, as shown in the figure. Editing a bride span in 3D is challenging so pan and zoom as necessary.**

7. **Turn off the Bridges layer, and save your project.**

Conduct a line-of-sight analysis

You can use lidar data to determine line-of-sight obstructions between observer points. This information can be useful for security or development purposes. There are many 3D tools for conducting visibility studies. You will use two, Construct Sight Lines and Line of Sight. In this exercise, you will use two observer points already created, one from the top of the US Steel building (Observer1) and the other at the fountain at Point State Park (Observer2).

1. Use the Line of Sight View bookmark for the 3D Scene and turn on the Observer1 and Observer2 layers.

2. On the Analysis tab, click the Tools button > Toolboxes tab, and expand 3D Analyst Tools > Visibility.

3. Click Construct Sight Lines, fill in the form as shown, and run the tool.
   - **Observer Points:** Observer1
   - **Target Features:** Observer2
   - **Output:** SightLine

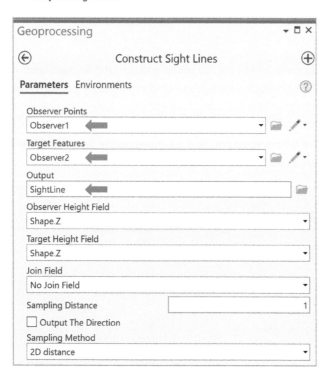

A sight line appears in the view between the two observer points.

4. Click the back button, and open the Line of Sight tool.

5. **Fill in the form as shown, run it, and close the tool when it finishes.**
   - **Input Surface:** Chapter11_LasDataset.lasd
   - **Input Line Features:** SightLine
   - **Output Feature Class:** LineOfSight

6. **Turn off the Chapter11_LasDataset.lasd layer and Z Mode to see the features that are visible (green) and not visible (red) between the observer points.**

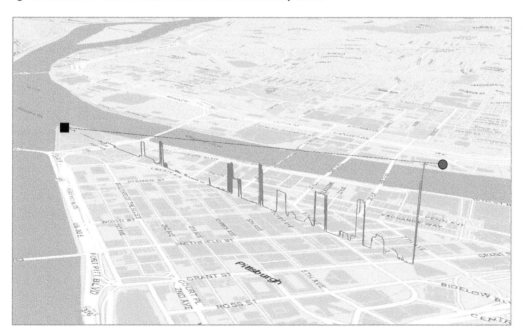

YOUR TURN

Turn on the 3D buildings layer, and symbolize with 50 percent transparency to better see the obstructions. Save your project.

Tutorial 11-5: Work with 3D features

This tutorial will explore a few of the many ArcGIS Pro 3D edit and create tools. Some edit functions work only with features that are z-enabled. If your features are not created as 3D features, you must convert them to 3D before editing using the geoprocessing tools in the 3D Analyst toolbox. In this tutorial, you will edit building polygons that are already 3D features to create multiple floors in a building and view floors using a range slider and manually edit polygons' heights using z-constraints.

Open the Tutorial 11-5 project with 3D building polygons

The buildings you will edit are the Allegheny County Courthouse and the old county jail, designed by architect H. H. Richardson.

1.   **Open Tutorial11-5.aprx from the Chapter11\Tutorials folder, and save it as** Tutorial11-5YourName .aprx.

2.   **Use the Courthouse bookmark.** The map zooms to the area surrounding the courthouse and the jail.

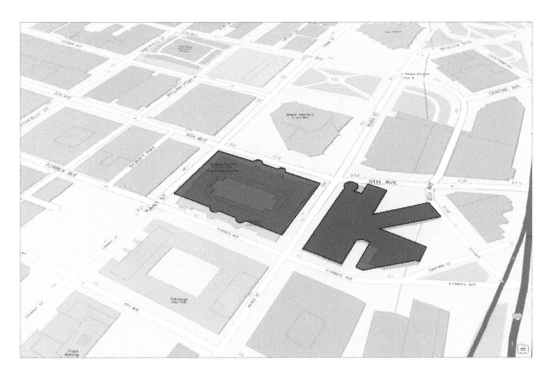

Extrude floors

First, you will create 3D floors for both the courthouse and the jail using the Duplicate Vertical tool. You can also use this tool to copy points or lines (for example, furniture or pipes) in a positive or negative direction if your features are above or below ground. You can also select and sketch on each new floor polygon.

1. On the Edit tab, in the Tools gallery, click the Duplicate Vertical button 🥩.

2. Using the Select tool, click the Allegheny County Courthouse polygon, and make the changes as shown.
    *   **Vertical Offset:** 20
    *   **Number of times to be duplicated:** 4

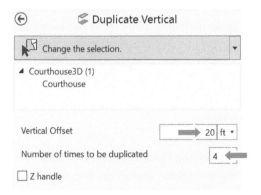

3.   **Click Duplicate.** There are five floors, and the Preview box will show the building floors as they are extruded.

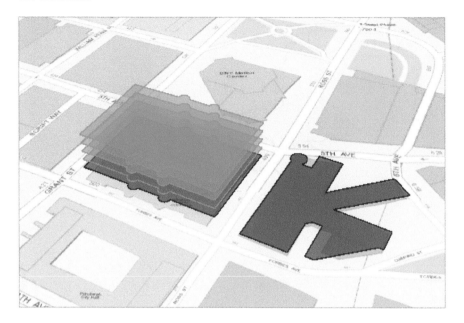

4.   **Clear the selected features, and save your edits.**

YOUR TURN

Use the Duplicate Vertical tool to extrude the floors of the old jail (building on the right), an offset of 20 feet, and use **3** for the number of times to duplicate. Clear your selections, save your edits, and close the Modify Features pane.

Use a range slider to view building floors

Setting range values offers a way to visualize certain floors in a building. This visualization method is especially useful if a floor contains detailed information or if a building has many floors. You can use this tool to visualize numeric values in an attribute table, including property values for parcels, crimes in neighborhoods, and so on.

1.   **Use the Building Floors bookmark.**

2.   **Open the Courthouse3D attribute table, and sort by Name.**

3.   **Select each floor of the courthouse, and in the corresponding FloorNumber field, type 1 for the first (lowest) floor, 2 for the second floor, 3 for the third floor, and so on. Repeat for the jail floors.**

4. Clear your selection, save your edits, and close the table.

5. In the Contents pane, right-click Courthouse3D, and click Properties. Under Range, click Add Range. For Field, make sure FloorNumber is selected, click Add, and click OK. A slider bar will be added.

6. Starting at the bottom, drag the slider up to 3, and notice floors 1 and 2 disappear. You can use range sliders in 3D animations, and you can modify range properties on the contextual map, on the Range tab.

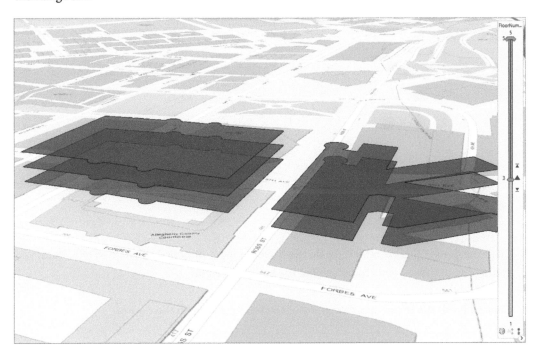

7. Drag the slider down to 1 to view all floors.

8. On the Range tab, in the Active Range group, click <None> for the Name value. This will turn off the range slider. You can also permanently remove the range by clicking Courthouse > Properties > Range.

Edit a building's height using dynamic constraints and the attribute table

Buildings are sometimes composed of multiple polygons at different heights. If these heights are not already derived from lidar data, you can use interactive handles to adjust the building height dynamically using a z-constraint or by typing the building height using attributes.

1. In the Contents pane, turn off Courthouse3D, and turn on Courthouse3DTowers.

2. **On the Edit tab, click the Modify button, and in the Modify Features pane under Alignment, click Scale.**

3. **Click the large tower on Grant Street.** The dynamic constraint icon will appear on the tower polygon. If your icon does not appear, you can turn on Show Dynamic Constraints In The Map by clicking Project > Options > Editing. You can adjust the map if necessary to better see the tower and constraint icon.

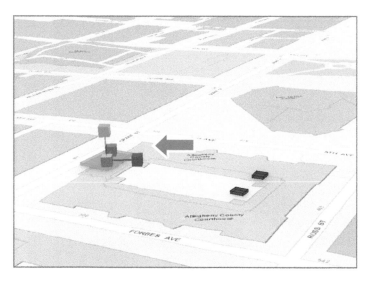

4. **Click the green (Z) constraint to scale the tower in the z-direction.** If you had lidar data, you could snap to those points to determine the building height.

5. **Click to finish the tower approximately at the height shown in the figure.**

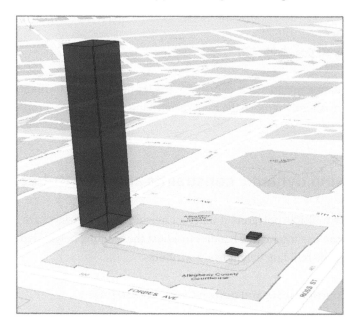

6. **Save your edits.**

> ## YOUR TURN
>
> An alternative to dragging features is to type the building heights in the corresponding attribute table. Open the Courthouse3DTowers attribute table, and type **150** for the two smaller towers' height. Clear your selected features, and save your edits and your project.

Tutorial 11-6: Use procedural rules and multipatch models

A CityEngine rule package (.rpk) is a compressed file that contains a compiled rule and all the assets (textures and 3D models) that the rule logic uses for creating 3D content. You can use these packages in ArcGIS Pro to create symbology that constructs and draws the procedural features on the fly from the source data. Another method creates 3D models and stores them as a feature class called a multipatch, whose features are a collection of patches that represent the boundary of a 3D object. A multipatch stores color, texture, transparency, and geometric data in its features.

Here, you will apply a predefined rule package to Pittsburgh's tallest office building, the US Steel building. You will also view multipatch features whose building facades were created in CityEngine using actual building facades. These features can take a long time to render, so it's recommended that you use a high-end graphics card and follow the hardware requirements.

Open the Tutorial 11-6 project with a building footprint

When you apply procedural rules, you must display features as layers in a 3D scene. The feature class polygon itself does not have to include z-values, but it must be in a 3D scene, and you can use 2D layers, such as building polygons.

1. **Open Tutorial11-6.aprx from the Chapter11\Tutorials folder, and save it as** Tutorial11-6YourName .aprx. The project opens with a 3D scene—US Steel Building—and one 2D building polygon footprint.

2. **Use the US Steel Building bookmark.**

Apply building rules using stacked blocks

You can apply a procedural rule to a building for a stacked block or more realistic high-rise or office building.

1. **In the Contents pane, click the red symbol for the US Steel building.**

2. **In the Symbology pane, click Gallery > All styles, and in the search box, type** Procedural, **and press Enter.** Procedural rules will update with new software releases and can be downloaded and added from CityEngine or ArcGIS Living Atlas of the World.

3. **Click the Stacked Blocks procedural symbol.**

4. **In the Symbology pane, click Properties, and click the Layers button** ⬚.

5. **Under Units, choose Feet; under Total Height, click the Container button** ⬚ **on the right of the current height. Select the Height field, and click OK.** This step sets the building height to a height field in which building heights are already entered.

6. **Fill in the form as shown, click Apply, and drag the USSteelBldg layer to 3D Layers.**
   • **Units:** Feet
   • **Total Height:** 40
   • **Levels:** 64

7. In the Contents pane, right-click USSteelBldg, and click Properties. On the Layer Properties dialog box, click Elevation, and under Features Are, click On The Ground, and click OK. The building will now be wrapped, showing the number of levels, or floors.

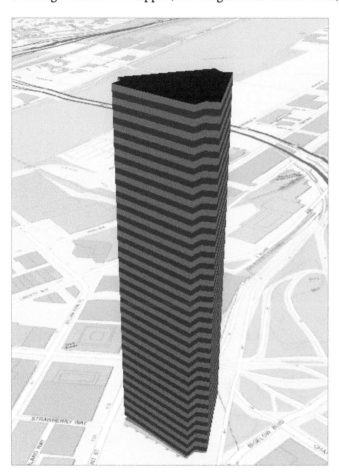

Apply an international building rule

The unit for international buildings is meters, so you will type the height instead of using the building height field, which is in feet.

1. In the Symbology pane, click Gallery, click the International Building procedural symbol, click Properties, and click the Layers button.

2. **Fill in the form as shown, and click Apply.**
    - **Units:** Meters
    - **BuildingType:** Highrise
    - **FloorHeightGround:** 7
    - **FloorHeightUpper:** 7
    - **Total Height:** 250

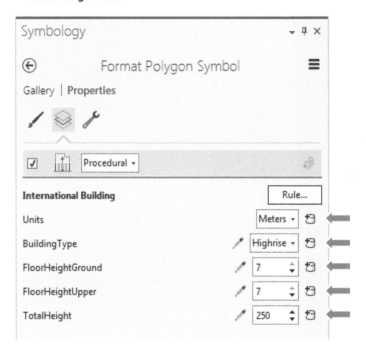

Although the result does not look exactly like the actual US Steel building, it is more realistic than a wrapped, level building.

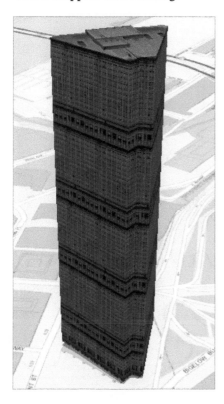

3.   **Close the Symbology pane.**

View multipatch models of buildings and street furniture

Smithfield Street is a study area in downtown Pittsburgh using multipatch layers of exported SketchUp Collada (.dae) files with realistic building images and street furniture. You will turn on these layers to explore multipatch models.

1.   **Close the US Steel Building scene, and open the Smithfield Street scene.**

2.   **Use the Smithfield Street bookmark.** Wait while the building facades display with textures.

3. **Turn on the Smithfield Street Furniture layer, change the basemap to Dark Gray Canvas, and
    wait for the view to render.** This layer of detailed features such as planters, garbage and recycling
    cans, and newsstands may take a while to render, depending on your graphics card.

YOUR TURN

Turn off the Smithfield Street Furniture layer, and turn on the Smithfield Textured 2 layer. View
the scene from various locations, including the Smithfield Street Bridge bookmark. Turn on the
Smithfield Buildings layer, and on the Appearance tab, in the Effects group, change the transparency to 10 percent. Save your project.

Tutorial 11-7: Create an animation

Animations are created by capturing an ordered set of viewpoints as keyframes and managing how the camera transitions between them. In this tutorial, you will take advantage of the bookmarks already in the scene to build a fly-through animation of downtown Pittsburgh. You will then improve the flight path and flight speed by manually inserting more keyframes and adjusting their timing.

Open the Tutorial 11-7 project, and explore bookmarks

1. **Open Tutorial11-7.aprx from the Chapter11\Tutorials folder, and save it as** Tutorial11-7YourName .aprx.

2. **On the Map tab, click Bookmarks > Manage Bookmarks, and in the Bookmarks pane, double-click each bookmark in order, from Frame1 through Frame8.** You will use these bookmarks to create the animation.

3. **Double-click the Frame1 bookmark.** This bookmark will be the first camera location of your animation.

Add an animation to the project, and create keyframes

Adding animation to a project enables the animation functions. As soon as you add an animation, you begin making keyframes manually. You can also import keyframes into an animation on the Animation tab, in the Create group, using the Import function.

1. **On the View tab, in the Animation group, click the Add button** 🖼. The Animation tab appears, and an Animation Timeline pane appears at the bottom of the screen. You are ready to make your first keyframe using bookmarks.

2. **In the Animation Timeline pane, click Create First Keyframe.** A thumbnail image of the starting location (Frame1 bookmark and the first keyframe) appears.

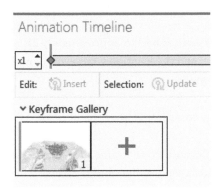

3. In the Bookmarks pane, double-click Frame2, and in the Animation Timeline, click the Append Next Keyframe button ⊞. This step adds the next keyframe to the animation.

4. Repeat step 3 for each of the remaining bookmarks, Frame3 through Frame8.

5. Close the Bookmarks pane. Now all the keyframes are added to the animation.

Play an animation, and change the duration

The keyframes have durations of three seconds between each frame for a total duration of 21 seconds. You can play the animation for this duration or extend the playback time.

1. On the Animation tab, in the Playback group, click the Play button ▶. On the Animation Timeline tab, the spacebar is a keyboard shortcut for Play/Pause.

2. On the Animation tab, in the Duration box, type 00:30.000.

3. Play the animation from the beginning. The animation will now play for 30 seconds.

Create a pause

Holding a keyframe will pause the animation at the selected frame. Here, you will add a hold to create a slight pause between frames 5 and 6 along the Smithfield Street Bridge.

1. On the Animation Timeline, double-click keyframe 5, and click the Hold button ⁺∥.

2. Play the animation from the beginning, and observe the slight pause between frames 5 and 6.

Add and delete keyframes

You can move keyframes to adjust the speed of animations and insert frames between keyframes using different camera locations. Here, you will add a new keyframe and manually adjust the camera.

1. On the Animation Timeline, drag the Time Indicator (red vertical bar) to approximately 21 seconds. Moving the Time Indicator scrubs through time, and the location where you move it is where you will adjust the camera and insert a new keyframe.

2. Using the wheel button or keyboard keys, adjust the view to a slightly lower focal point.

3. On the Animation timeline, click the Update button 🔄. Because no keyframe existed at 21 seconds, when you click update, it will insert a new keyframe to capture the camera change and add a new keyframe using the current camera location.

4.   Play the animation from the beginning.

View an animation's path, and manually edit a keyframe's properties

When adjusting animations, it's useful to see the camera's path and keyframes so that you can move to a keyframe and adjust its properties, such as z-elevation (camera location).

1.   On the Animation Timeline, double-click frame 1.

2.   From Contents, right-click Animation Scene, and click Properties. Click Extent > Use Extent Of Data In All Layers, and click OK. Some keyframe locations would not be visible because of the extent setting. The next step will show these keyframe camera locations.

3.   On the Animation tab, in the Display group, click the Path button ![icon]. This step turns on the camera path and keyframes to better see how the animation plays. Keyframe 2 is a little too high, so you will edit its properties to lower the keyframe.

4.   On the Animation Timeline, in the Keyframe Gallery, double-click keyframe 2.

5.   On the Animation tab, in the Edit group, click the Properties button ![icon], expand Camera, and type 1000 for the z-value.

6.   Close the Animation Properties pane, and click the Path button to turn off the path.

7.   Play the animation from the beginning. Notice that the camera is lower at keyframe 2.

Create a movie from the animation

Now that you have created an interesting animation, it's time to share the animation. To share the animation, you will export a movie to a file. You have several options, including exporting your movie directly to YouTube, Vimeo, and so on, or as a draft animation. You can change the file location and movie resolution, type, size, and so on in the File Export and Advanced Movie settings.

1.   On the Animation tab, in the Export group, click the Movie button ![icon].

2.   In the Export Movie pane, in the Movie Export Presets group, click the Draft button. This step creates a smaller file. The resulting quality of the file isn't high, but it is much faster to produce.

3.   Under File Name, click the Browse button, browse to Chapter11\Tutorials, and type Animation3D as the movie name.

4. **In the Export Movie pane, click Export. Wait while the movie is created.** If you use other settings with higher resolutions or a larger size, the movie can take a long time to render. You can save the media format as separate.jpg files that you could later stitch together using another animation software application.

5. **In File Explorer, browse to Chapter 11\Tutorials, and double-click Chapter11Animation3D.mp4 to play the movie.**

6. **Save your project.**

Assignments

This chapter has two assignments to complete that you can download from ArcGIS Online, at go.esri.com/GISTforPro2.8Data:

- **Assignment 11-1**: Prepare 3D building and topography features for a 3D study.
- **Assignment 11-2**: Create a realistic 3D scene for a campus study.

PART 4

Managing operational systems with GIS

Operations management with GIS

Graffiti Mapping System

LEARNING GOALS

- Get an introduction to operations management systems.
- Create tasks to prepare data for an operations management system.
- Build a ModelBuilder model to be used in a task.
- Share web layers in ArcGIS Online for use in an ArcGIS® Dashboards operation view.
- Create an ArcGIS Online map for use in a Dashboards operation view.
- Create and use a Dashboards operation view.

Introduction

Up to this point, this book has dealt primarily with GIS applied to one-time projects in which spatial data is collected at one point in time and used in GIS to help solve a problem or shed light on an issue. For instance, the GIS in chapter 1 addressed the issue of geographic access to urgent health care clinics in parts of Pittsburgh with low-income populations. Although the GIS could be updated with new data over time, no organization or system was envisioned to continue the work.

In contrast with a one-time project, operations management applies to organizations that deliver goods and services. Operations signifies continuous work to provide goods and services to customers. Such work generates and depends on demand data for goods and services. Management refers to the oversight function for efficient, timely, and responsive delivery of demanded services. An operations management system is an information system with supporting computer applications. A significant feature of such systems is that they process and use a continuous flow of demand data, in contrast with projects and one-time data collection. Because demand data is often spatial with point locations of demands, operations management systems often include GIS.

In this chapter and chapter 13, you will build two GIS-based operations management systems for graffiti. In this chapter, you build the Graffiti Mapping System to help police identify patterns of serial graffiti artists, apprehend the artists, and prevent future graffiti. The operations management system you'll build in chapter 13, the Graffiti Removal System, is for the public works employees who remove graffiti from public surfaces, such as buildings, walls, and bridges. The system performs optimal routing for graffiti removal at several locations during a work shift. Optimal routing determines the sequence of stops at graffiti sites to minimize travel time.

In this chapter, you'll create tasks, a ModelBuilder model, and a Dashboards operation view that police will use as their main tool to prevent graffiti. First, it may be helpful to review the following information:

- Tasks are stored, reusable workflow guides that load commands and manage map views and layers, thereby reducing mouse clicks and overhead. Task processing and output are consistent and reliable, regardless of who does the work, because of the prepackaging and instructional steps of tasks.
- Tutorial 10-3 introduced ModelBuilder and models. A model automates a workflow consisting of two or more geoprocessing or analysis tools that become the processes of the model. Each tool or process is an algorithm that transforms inputs into outputs. Processes are strung together, with the output of one process becoming the input of another process. Therefore, instead of searching for and running tools one after the other on your own, you run a sequence of tools with one click of a button in a model.
- ArcGIS Dashboards is an app for monitoring and reporting operations data over time. An operation view has one or more maps, plus additional data displays such as statistics, charts, and queries.

The system in this chapter and the one in chapter 13 use 311 data as inputs. Increasingly, citizens of American cities request nonemergency public services using the phone number 311, which is also the name of the one-stop, gateway department for providing such services. Besides calling 311, citizens can often use a 311 website to record calls for service. The 311 system automatically sends service calls to responsible city departments such as police, public works, building inspection, and animal control to provide requested services. The reports of citizens asking for graffiti removal are a major type of 311 data. This chapter uses 311 graffiti call data from Pittsburgh, but specific graffiti locations are random to protect data privacy.

The Graffiti Mapping System in this chapter is fictitious and not a real police system. Nevertheless, the system would be relevant for the Pittsburgh Police's Graffiti Task Force, which targets serial graffiti artists that cause damage to public and private property or paint racist or other hate graffiti.

Suppose that police determine that 12 weeks (84 days) of data is enough to identify serial graffiti artists. For raw data input to the Graffiti Mapping System, assume that the Pittsburgh 311 system exports a CSV file with the most recent 84 days of graffiti data each time you need to produce a weekly map.

Included in the CSV data are the longitude and latitude of each graffiti location, shown as X and Y, respectively, in the sample data that follows. Graffiti artists are not known by their actual names, but rather by monikers such as Phase III or Tommy T, as seen in the Artist attribute. Graffiti_Type_Code has code values Tag, Throw Up, and Piece, which are standard categories of graffiti. Tag is the simplest graffiti, with just the sign or moniker of a graffiti artist, generally with only one color. Throw Up is quickly painted graffiti with some graphic features in addition to a tag and generally with a few colors. Piece is complex graffiti, a "piece of art," with several colors.

REQUEST_ID	ADDRESS	ZIPCODE	DATE_CALL	GRAFFITI_TYPE_CODE	ARTIST	X	Y
72024	642 W NORTH AVE	15212	5/23/2016	Piece	Phase III	-80.0132	40.45445
73603	432 AVERY ST	15212	5/27/2016	Piece	Phase III	-80.0007	40.45329
59024	3803 BATES ST	15213	4/4/2016	Tag	Tommy T	-79.9533	40.43872
61138	3564 BLVD OF THE ALI	15213	4/14/2016	Tag	Tommy T	-79.9546	40.43475
61229	3401 JUNO ST	15213	4/14/2016	Tag	Tommy T	-79.9513	40.43504
61230	362 YORK WAY	15213	4/14/2016	Tag	Tommy T	-79.955	40.43844
54525	164 N SHORE DR	15212	3/12/2016	Tag		-80.0131	40.44507
54856	1201 ELWELL ST	15207	3/14/2016	Tag		-79.9153	40.36946
54892	1568 RIVER AVE	15212	3/14/2016	Throw Up		-79.9831	40.45999

Tutorial 12-1: Create tasks for the Graffiti Mapping System

A task item is a container that has a set of related tasks. Each task has one or more steps.

You'll create a task item named Prepare Graffiti Data for ArcGIS Online for the weekly preparation of data for the Graffiti Mapping System. The task item has two tasks, Prepare Data and Publish Web Layers, and the following multiple steps in ArcGIS Pro.

The **Prepare Data** task has steps to enter a date in a feature class named Heading and create a feature class named Graffiti. Heading is a point feature class with a single point. The point has a date attribute, also named Heading, which stores the end date of the 84-day time window of graffiti data to be mapped. Every week, when you produce a new graffiti map, you must edit this date to be the current end date of the time window. The dashboard you'll build will display this date so that the user can learn the vintage of the map's data. The second step creates and symbolizes the Graffiti feature class from the 84-day XY data table provided by the city's 311 system.

Prepare data

1. The **Enter Date** step opens the Heading attribute table and guides the user in editing the Heading date value.
2. The **Save Edits** step saves the edited date.
3. The **Create Graffiti Feature Class** step runs a ModelBuilder model named Graffiti Data Import that you'll build in this tutorial. The model adds raw CSV data as an XY event layer to the ArcGIS project, creates the Graffiti feature class with projected Web Mercator coordinates from the XY data, and symbolizes the Graffiti layer using unique values by graffiti type.

Publish web layers

1. The **Share Web Layers** step publishes the Heading and Graffiti web layers to your ArcGIS Online account. After you publish the web layers and they are added to an ArcGIS Online map, you can incorporate them into an operations view for analysis by police and other city officials who work on graffiti prevention.
2. The **Share Heading Web Layer** step publishes the Heading web layer.
3. The **Share Graffiti Web Layer** step publishes the Graffiti web layer.

You'll build the model of the Create Graffiti Feature Class step. Then you can create the tasks and steps of the task item with no interruptions.

Open the Tutorial 12-1 project

1. **Open Tutorial12-1 from Chapter12\Tutorials, and save it as** Tutorial12-1YourName.

2. **Use the Pittsburgh bookmark.** You'll build a ModelBuilder model in this exercise for creating and symbolizing a Graffiti feature class from XY event data. When you create a Graffiti feature class, it will replace the starting Graffiti feature class and symbolize it using unique values for types of graffiti. The starting Graffiti layer in the map is symbolized with a single symbol so that you can see that the model creates a Graffiti feature class and changes symbology to unique values.

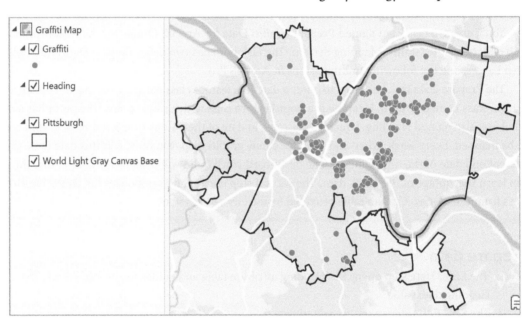

Build the Graffiti Data Import model

When you build a model, you search for geoprocessing tools and add them to the model. Once you connect data to the tools and set tool parameters, you'll have model processes. The model you'll build in this exercise, as shown in the figure, has three processes:

*   The first process imports the raw XY graffiti data, which has geographic latitude-longitude coordinates, into a feature class with the same coordinates.
*   The second process projects the feature class from geographic coordinates to Web Mercator coordinates that are preferred for ArcGIS Online.
*   The third process symbolizes the map layer with unique symbols for type of graffiti.

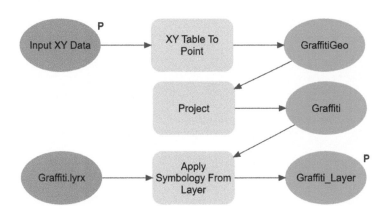

1.   On the Analysis tab, in the Geoprocessing group, click the ModelBuilder button.

2.   On the Analysis tab, in the Geoprocessing group, click the Tools button, and in the Geoprocessing pane, search for (but do not open) the XY Table To Point tool.

3.   Drag the XY Table To Point tool to the Model window.

4.   **Similarly, search for and add the** Project **and** Apply Symbology From Layer **tools to the Model window in order, as shown in the figure, and when finished, close the Geoprocessing pane.**

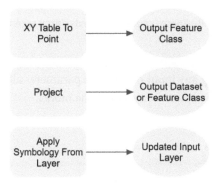

Configure the XY Table To Point process

1.   In your model, double-click the XY Table To Point tool element.

2.   In the window that appears, for Input Table, browse to Chapter12\Data and double-click **Graffiti20160603.txt**. Graffiti20160603.txt is the raw CSV data from a 311 system. ArcGIS Pro automatically identifies and adds the X and Y fields, X and Y, as well as the correct spatial reference of GCS_WGS_1984 (geographic, latitude and longitude). Note that Spatial Reference is for the current coordinates of the XY data.

3.   **Rename Output Feature Class** GraffitiGeo.

4.   **Click OK.** The process and its input and output diagram elements now have color fill, indicating that the process is configured, valid, and ready to use.

5.   **Right-click the input ellipse, Graffiti20160603.txt, click Rename, replace the existing name by typing** Input XY Data, **and press Enter.**

6.   **Right-click the input again, and click Parameter.** Making any input or output a parameter allows you to search for the input data or name and locate the output when you run the model.

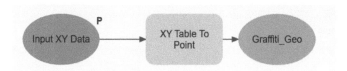

Configure the Project process

1.   **Double-click the Project process.**

2.   **Make selections or type as shown in the figure.**
   - **Input Dataset or Feature Class:** GraffitiGeo
   - **Output Dataset or Feature Class:** Graffiti
   - **Leave Input Coordinate System and Geographic Transformation blank.**
   - **For Output Coordinate System, click the Select Coordinate System button 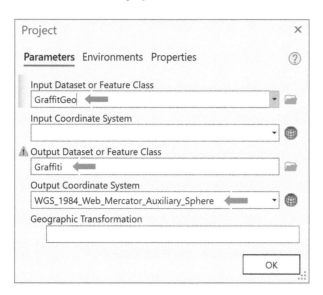, expand the Coordinate System tree to get Projected Coordinate System > World > WGS 1984 World Mercator (Auxiliary Sphere).**

3.   **Click OK to complete the tool settings.**

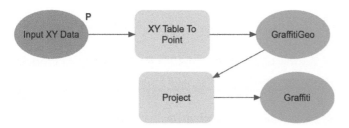

Configure the Apply Symbology From Layer process

1.   To connect Graffiti to Apply Symbology From Layer, click Graffiti and drag the resulting arrow, connect the arrow to Apply Symbology From Layer, and select Input Layer.

2. **Double-click the Apply Symbology From Layer process, and for the Symbology layer, browse to Chapter 12\Tutorials\Resources, select Graffiti.lyrx, make additional selections if necessary, and click OK.**
   - **Input Layer: Graffiti2**
   - **Symbology Layer: Graffiti.lyrx**
   - **Type: Value Field**
   - **Source Field: GRAFFITI_TYPE_CODE**
   - **Target Field: GRAFFITI_TYPE_CODE**

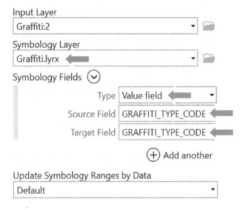

3. **Right-click the output of this process, Graffiti_Layer, and click Parameter.**

4. **Save and close your model.**

Rename and run the model

1. **Open Catalog, expand Toolboxes > Tutorial12.tbx, and right-click Model and select Properties. Set the Name as** GraffitiDataImport **and the Label as** Graffiti Data Import. **Click OK.**

2. **Right-click Graffiti Data Import, and click Edit.** The model opens in edit mode.

3. **On the ModelBuilder tab in the Run group, click Run, and after the model runs, close the window with processing information.** After the model runs, the old Graffiti layer is replaced by the newly imported and projected Graffiti layer, and the new layer is symbolized with unique symbols by graffiti type. Also, the process elements in the model have drop shadows, indicating that they have been run. If you need to run the model again in edit mode, you must click the Validate button in the ModelBuilder Run group, which removes the drop shadows. If you are diagnosing a model, you can run it step by step by right-clicking each process in turn and clicking Run. If there were any errors, they would be in red font in the processing window and would include diagnostic information.

4. **Right-click the output of the Apply Symbology From Layer process, Graffiti_Layer, click Add To Display, and look at the Graffiti map to see that it has new data and is symbolized.**

5. **Save and close the Graffiti Data Import model.**

6. **On the Analysis tab, click Tools, search for Graffiti Data Import, and click to open.** In this model interface, the user can browse for the needed graffiti XY data. Note that by naming the input file with the date in YYYYMMDD format, such as Graffiti20160603, the files are sorted chronologically.

7. **Click Run, and when the model finishes, close the Geoprocessing pane.** Note that this step over-writes the previously existing graffiti feature layer, which is the desired result. The destination dashboard for the graffiti feature class expects an updated feature class each week, rather than updating previous data by dropping the oldest week and adding a new week itself.

8. **Save your project.**

Tutorial 12-2: Create tasks to import graffiti data

Next, you will create the Prepare Graffiti Data for ArcGIS Online task item, its Prepare Data task, and the task's three steps, Enter Date, Save Edits, and Create Graffiti Feature Class. The task item has a second task, Publish Web Layers, which you'll create after this exercise. You must have completed tutorial 12-1 and built its Graffiti Data Import model for use in the Create Graffiti Feature Class step.

Open the Tutorial 12-2 project

1. **Open Tutorial12-2 from Chapter12\Tutorials, and save it as** Tutorial12-2YourName.

2. **Use the Pittsburgh bookmark.**

Create a task item for preparing graffiti data

1. On the Insert tab, in the Project group, click the Task button 🔲, and click **New Task Item**. The Tasks and Task Designer panes appear. You create task items, tasks, and steps in Tasks and configure them in Task Designer.

2. **If the panes are docked, undock them, and move them so that you can see the Contents pane and map.**

3. **Type the following information as shown in the text boxes in the Task Designer pane for the task item. Type your name for Author.** Note that the document Tasks_Copy_And_Paste.docx in Chapter12\Tutorials\Resources has much of the text that you are instructed to type. So instead of typing, you can open the document and copy and paste to save time.
   - **Name:** Prepare Graffiti Data for ArcGIS Online
   - **Author:** Mary Smith
   - **Summary:** These tasks prepare and share data on ArcGIS Online for use in a dashboard.
   - **Description:** Has tasks

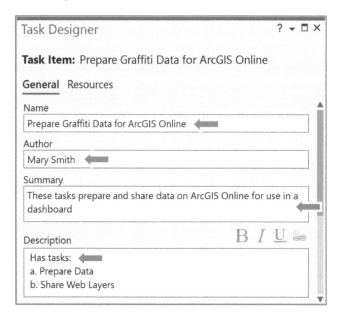

4. **Save your project.** Saving the project also saves the task item.

Create the Prepare Data task

1. **In the Tasks pane, click the New Task button 🔲.**

2.   **In the Task Designer pane for Name, type** a. Prepare Data, **and for Description, type** Has steps: 1. Enter Date 2. Save Edits 3. Create Graffiti Feature Class. **Ensure that you press Enter after each line of the description to arrange the description as shown in the figure.**

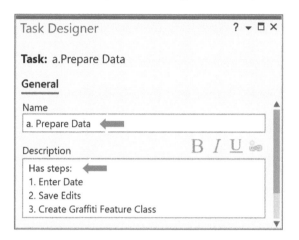

3.   **Save your project.**

Create the Enter Date step

1.   **In the Tasks pane, click the New Step button** 📋.

2.   **In the Task Designer pane, for Name, type** Enter Date.

3.   **For the step instructions, type** i. Enter a new date to replace the old date in the Heading field in the Heading attribute table. ii. Press Enter to update the Attribute field. iii. Click Next Step. **Ensure that you press Enter after each instruction as needed to place the start of the next instruction on a new line.**

4.   **For Step Behavior, click Auto Run.**

5.   **In the Task Designer pane, in the top horizontal menu, click Actions, point to the command bar, and click the Record button, as shown in the figure.**

The following step 7 will be recorded for automatic use when you run the task later. Notice that after clicking Record, when you point to any open user interfaces, the pointer gets a red circle with a center dot indicating that it's ready to record a command. After recording a single step (use of a command that accomplishes something), the pointer will revert to its original form.

6.   **Move the Tasks and Task Designer panes so you can see the ArcGIS Pro window.**

7. **In the Contents pane, right-click Heading, and click Attribute Table.** The pointer reverts to its original form, indicating that recording is finished with a single command (Attribute Table) recorded.

8. **In the existing row, under Heading, type** End Date: June 10, 2016, **and press Enter.**

9. **In Task Designer, click the Views tab, and ensure that Graffiti Map is active and open.**

10. **In Task Designer, click the Contents tab. Click at the intersection of Heading and Select Layer to turn on layer selection for the Heading layer.** This step ensures that the Heading layer is selected in the Contents pane so that the open attribute table command is applied to that layer.

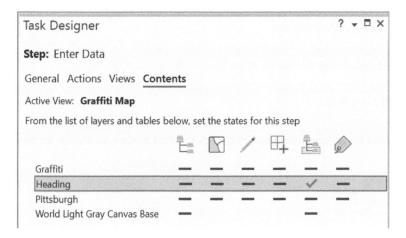

11. **Save the project.**

Create the Save Edits step

1. In the Tasks pane, click the New Step button.

2. **For Name, type** Save Edits; **for Instructions, type** Click Yes and close the table; **and for Step Behavior, select Automatic.**

3. **Click Actions, point to the command bar, and click the Record button.**

4. **In the ArcGIS Pro window, click the Edit tab, click Save, and click Yes.**

5. **Close the Heading table, and save your project.**

> ## YOUR TURN
>
> Try out your new task. In the Tasks pane, click the Options button, and click Exit Designer. In the Tasks pane, point to the Prepare Data task, click the blue arrow that appears on the right, and follow your instructions to edit and save the date, entering **June 12, 2016,** and pressing Enter for the new date.

Create the Add Graffiti XY Event Data step

Adding this step will result in creating a Graffiti feature class.

1. **Click the Options button in the Tasks pane, and click Edit In Designer.**

2. **In the Tasks pane, click the New Step button.**

3. **In Task Designer, on the General tab, for Name, type** Add Graffiti XY Event Data**.**

4. **For Step Behavior, select Auto Run.**

5. **Click the Actions tab, point to the command bar, and click the Edit button, as shown in the figure.**

6. **For Type of Command, select Geoprocessing Tool, search for and select** Graffiti Data Import, **and click OK.** The Embed portion of the task step exposes the model's parameter, the input x,y data table, so that the user can browse for the current data.

7. **Click Done, and save your project.** Note that if your project is closed at some point, you can open it in edit mode from the Catalog pane under Tasks.

> ## YOUR TURN
>
> Repeat all the steps of the Prepare Data task. Edit the date to **June 10, 2016**, and save the edits. The Add Graffiti XY Event Data step runs automatically, resulting in a new Graffiti feature class in the Contents pane. Click Finish.

Share the Graffiti and Heading web layers

The key to building the operation view in Dashboards later in this chapter is to keep the names of input map layers the same from week to week and run to run. You'll use Graffiti followed by Your Name for the fixed name of the graffiti map layer and Heading followed by Your Name for the layer with the current date. Then every time you open a corresponding map in ArcGIS Online, it will open with new data but with the same name over time, so you won't have to modify the map. The first time that you share (publish) Graffiti Your Name and Heading Your Name in ArcGIS Online, the command to use is Share As Web Layer. The second time you share these layers, and every time after that, the command is Overwrite Web Layer, which allows you to overwrite the previous versions with the new ones of the same name. So as a preliminary step, next you will use Share As Web Layer interactively. Then when you create a corresponding task, you'll use Overwrite Web Layer.

1. **Save your project, and close the Tasks pane.**

2. **In the Contents pane, right-click Graffiti_Layer, and click Sharing > Share As Web Layer.**

3. **In the Share As Web Layer pane, for Name, type your name at the end of Graffiti (for example,** Graffiti Mary Smith)**; for Summary, type** Twelve weeks of graffiti data; **for Tags, type** Pittsburgh **and** Graffiti.

4. **Click Publish.** Wait a minute or so until the web layer publishes.

5. **Likewise, share Heading as a web layer, add your name to Name (for example,** Heading Mary Smith)**, and for Summary, type** Stores end date of graffiti data, **and for Tags:** Pittsburgh **and** Graffiti Date.

6. **Finally, share the Pittsburgh layer by adding your name to Pittsburgh (for example,** Pittsburgh Mary Smith)**. Then for Summary, type** Pittsburgh's boundary, **and for Tags:** Pittsburgh **and** Boundary.

Create the Publish Heading and Graffiti Web Layers task

Currently, the Tasks and Task Designer panes should be closed. You can open or edit tasks from the Catalog pane, as you'll see next.

1. **Open the Catalog pane, expand Tasks, right-click Prepare Graffiti Data for ArcGIS Online, and click Edit In Designer.**

2. **Click the New Task button.**

3. **In Task Designer, for Name, type** b. Share Web Layers. **For Description, type** Has Steps: 1. Share Heading Web Layer 2. Share Graffiti Web Layer.

4. **Save your project.**

Create the Share Heading and Graffiti Web Layers step

1. **In the Tasks pane, with the Share Web Layers task selected, click the New Step button.**

2. **In Task Designer, for Name, type** Share Heading Web Layer. **For Instructions, type** i. In the Contents pane, right-click Heading and click Sharing > Overwrite Web Layer. ii. Click My Content, select Heading Your Name, and click OK. iii. In the warning window, click OK. iv. Click Publish. v. When publishing finishes, close the Overwrite web layer window.

3. **For Step behavior, select Auto Run.**

4. **In the Tasks pane, click the back arrow button.**

5. **Click the New Step button.**

6. **In Task Designer, for Name, type** Share Graffiti_Layer Web Layer. **For Instructions, type** i. In the Contents pane, right-click Graffiti_Layer and click Sharing > Overwrite Web Layer. ii. Click My Content, select Graffiti Your Name, and click OK. iii. In the warning window, click OK. iv. Click Publish. v. When publishing finishes, close the Overwrite web layer window.

7. **For Step behavior, select Auto Run.**

8. **Save your project.**

> ## YOUR TURN
>
> Repeat both tasks, Prepare Data and Share Web Layers, of the Prepare Graffiti Data for ArcGIS Online task item. First, resymbolize Graffiti with a single symbol so that you can see when the Create Graffiti Feature Class step of the Prepare Data task runs, creating Graffiti again and symbolizing it with unique values. In the Enter Data step, enter the end date of **June 3, 2016**. When you finish, save your project, and close ArcGIS Pro.

Tutorial 12-3: Create a map for Dashboards

You'll create a map in ArcGIS Online that will be reused each time new Graffiti and Heading web layers are shared. To run this tutorial, you must have the web layers from tutorial 12-2 published to your ArcGIS Online account. You'll add the published web layers to a map and create pop-up windows and bookmarks.

Add layers to the map

1. **Go to ArcGIS.com, and sign in to your organizational account.**

2. **Click Map on the main toolbar.**

3. **Click the Layers button > Add Layer.**

4. **In the Add Layer panel, click the Add button next to Graffiti Your Name to add it.**

5. **Follow the same operation to add the layers Heading Your Name and Pittsburgh Your Name, and click the back arrow next to Add Layer.**

6. **Click the three dots on the right of each layer to rename them** Graffiti, Heading, **and** Pittsburgh.

7. **In the Layers panel, drag the layers to be ordered from top to bottom, Graffiti, Pittsburgh, Heading.**

8. **Click Heading, and in the right menu under Properties, disable the legend. Do the same for Pittsburgh.**

9. **Click the Basemap button, and change the basemap to Light Gray Canvas.** If Light Gray Canvas is not available, select another simple basemap that has streets when zoomed in.

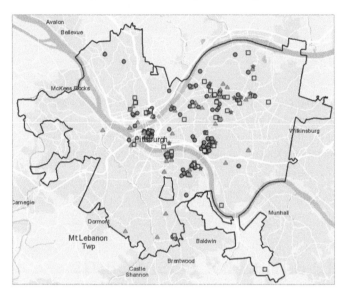

10. **Save the map as** Graffiti Map Your Name **with the tags** Pittsburgh **and** Graffiti, **and the summary of** Current graffiti map of Pittsburgh.

Add and configure pop-up windows

Dashboards uses the pop-up windows that you will configure in your map.

1. **In the Layers pane, click Graffiti, and on the right-hand settings toolbar, click Configure Pop-ups.**

2. **For Title, type** Graffiti, **and click OK.**

3. **Click Fields List, and click Select Fields.**

4. **Click Select All, and click it again to Deselect All.**

5. Click to turn on the following attributes: DATE_CALL, ARTIST, ADDRESS, ZIPCODE, GRAFFITI_ TYPE_CODE, and COMMENT. Click Done.

6. Using the six dots on the left of an attribute, drag COMMENT up to follow ZIPCODE.

7. In the Layers panel, click Pittsburgh and disable its pop-ups. Do the same for Heading.

8. Save the map.

YOUR TURN

Click a graffiti point on the map to try out your pop-up window.

Create bookmarks

Dashboards uses the bookmarks that you configure in ArcGIS Online.

1. Zoom in or out so that Pittsburgh fills the map.

2. Click the Bookmarks button, and click Add Bookmark. Type Pittsburgh for Title, and click Add.

3. Zoom in to the Central Business District of Pittsburgh.

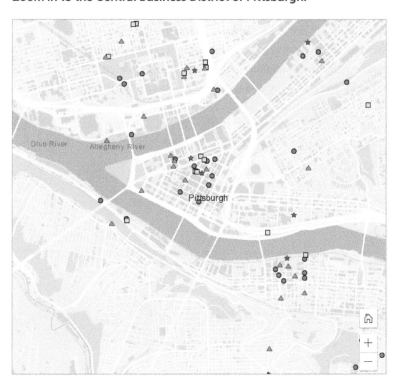

4.   **Add a bookmark titled** Central Business District.

5.   **Use the Pittsburgh bookmark.**

6.   **Save your map, and leave ArcGIS Online open.**

Tutorial 12-4: Create a graffiti dashboard using Dashboards

A dashboard is an app intended for an organization to use for operations management, providing multiple up-to-date visual displays, data, and statistics for understanding demand and supply patterns for goods or services or other mission-critical data. A dashboard can display an ArcGIS Online map that you build (such as in tutorial 12-3), attribute data lists, various charts and descriptive statistics, and queries for which the user can choose conditions.

The dashboard you will build in this tutorial is for identifying and investigating the patterns of serial graffiti artists for graffiti prevention and apprehension of graffiti artists by police. With a serial artist's spatial and temporal patterns understood, police can target patrols to specific areas to raise the risk of arrest and thereby prevent graffiti or apprehend the artist in the act of painting graffiti.

The following figure shows the finished graffiti dashboard as it appears when open. The map displays 12 weeks of the most recent graffiti data (with end date June 3, 2016, in this tutorial), symbolized by type of graffiti. If you click a graffiti symbol on the map, you get the point's pop-up with data. Also available in the upper-right menu of the map, besides the usual map navigation of panning and zooming, are search, the map legend, and the ability to turn layers on and off.

The side panel allows you to limit the display to any of the graffiti artists active during the past 12 weeks who left their tag names on graffiti. The default is that no artist is selected (None) so that all data displays. The top of the middle panel has the end date of the data, from the Heading web layer. The bar chart shows the frequency distribution of graffiti calls by type for the current map display, and the list is the corresponding data sorted descending by date of call and then ascending by address. If you click a row in the list, the map zooms and pans to the corresponding point and flashes the point.

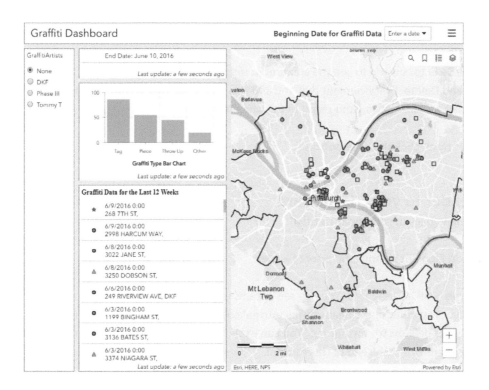

Create a dashboard

1. In the upper right of your screen, click the App Launcher button ⠿, and click **Dashboards**.

2. **Click Create Dashboard. For Title, type** Graffiti Dashboard Your Name; **for Tags, type** Pittsburgh and Graffiti; **and for Summary, type** Displays graffiti map, data, and queries for the last 12 weeks.

3. **Click Create Dashboard.** ArcGIS Online creates a blank dashboard in edit mode, ready for you to configure with resources from your ArcGIS Online web maps.

Add a web map to your dashboard

1. **Click the Add Element button** ⊞▽, **select Map, and click Graffiti Map Your Name.** That action opens the Settings panel for map properties as an element of the dashboard.

2. **In Settings, select Ruler for Scalebar, and enable all except Basemap Switcher.** These selections populate a toolbar for the map in the dashboard and create a scale bar.

3. **Click the General tab, change the Name to** Graffiti, **and click Done.** Your map is added as the dashboard's only element along with its toolbar in the map's upper-right corner and scale bar in the lower-left corner.

4. **Point to the map and notice the minimized, blue Selector button in the upper-left corner.**

5. **Point to the Selector button to see the element items menu, which has Drag Item, Configure, Duplicate, and Delete.** Every element that you add to a dashboard (for example, map, query, list, or chart) has a Selector button and menu items when in edit mode.

6. **In the upper right, click the Save button.**

Add the header to your dashboard

The header element gives the dashboard a title.

1. **Click the Add Element button, and select Header.**

2. **For Title, type** Graffiti Dashboard. You can see options to format elements, such as changing the background or text color of the header. This tutorial covers only the basics of building a dashboard, using the defaults for formatting. You can explore formatting dashboard elements in a Your Turn assignment at the end of this tutorial.

3. **Click Done.** The title appears at the top of the dashboard.

Add a side panel to your dashboard

1. **Click the Add Element button, and select Side Panel.** The side panel has the purpose of containing one or more queries that the user can execute. Your dashboard will have a single query to display the graffiti of an artist, as identified by their tag.

2. **In the Appearance panel, for Title, click the Edit button, and type** Graffiti Artists. Clicking an edit button exposes the rich text editor, with which you can format text, add an image or table, and add a hyperlink. Note that the header and side panel elements are the only two elements of the dashboard for which you cannot change location. You can change the height of the header to small, medium, or large, but the width of the side panel cannot be changed. You can reposition and size all other elements.

3. **Click Done.**

4. **Point to the side panel's Selector button, and click the Add Category Selector button.**

5. **For Categories From, select Grouped Values.** This option forms groups using repeating values in data records such as a graffiti artist name.

6. **For Layer, click Graffiti.**

7. **Type or make selections as shown to complete the selector.** Also, for Preferred Display Type, scroll down after clicking the drop-down menu to find option buttons.
   - **Category Field:** Artist
   - **Preferred Display Type: Radio buttons**
   - **None Option: On**
   - **Name:** Artist Selector

Selector options   Show data table

Data

Categories from   Defined values | Features | Grouped values

Layer: Graffiti   Change

Filter   + Filter

Category field   ARTIST ◄━━━   ▼

Category   Label
No override defined

+ override | Load categories

Maximum categories   50

Sort by   Add field ▼
   ARTIST   ⬆ ⬇ 🗑

Selector

Label

Selection   Single | Multiple

Operator   equal   ▼

Preferred display type   Radio buttons ◄━━   ▼

Display type threshold   10

None option   ⬤━ ◄━━

Placement   First | Last

Label for none   None

Default selection First | Last

General

Name   Artist Selector ◄━━

No data   Default

No selection   Default

8. **In the Category Selector panel on the upper left, click Actions.**

9. **Click Add Action > Filter.**

10. **Click Add Target > Graffiti, and click Done.** This action causes the Category Selector to apply its condition, the name of an artist, to the map and filter graffiti points to show only that artist.

11. **Click the option button for artist DKF, and click the southernmost graffiti point to see its pop-up.**

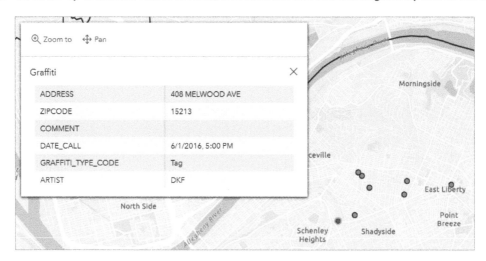

12. **Close the pop-up, and in the side panel, click None to restore all graffiti points to the map.**

Add a list to display the ending date of the graffiti data

The list will display the only row of the Heading web map layer, which has the ending date of the 12 weeks of graffiti data. This date is important for informing the user that the dashboard's data is up-to-date.

1. **Click the Add Element button, select List, and for layer, select Heading.**

2. **In the Data Options panel, click the General tab.**

3. **In General Options, for Name, type** Ending Date List.

4. **Click Done.**

5. **Point between the Ending Date List and the map so that you get the double-headed arrow for changing column width, and make the list column as narrow as possible so that its text, End Date: June 3, 2016, remains on one line.** Depending on how you completed tutorial 12-3, your end date may vary. You'll place another list for graffiti data below the heading in the next exercise. You'll also place a bar chart for frequency of graffiti types in the same column.

6. **Save your dashboard.**

Add a list to display graffiti data

1. **Click the Add Element button, select List, and for layer, select Graffiti.**

2. **For Maximum Features Displayed, type** 100.

3. **For Sort By, click Add Field and select DATE_CALL. Set it to Sort Descending.**

4. **Click Add Field again, and select ADDRESS.** Graffiti data will sort first by the date of the 311 call descending (newest first), and then within date group by address.

5. **Click the General tab for General Options, and for Name, type** Graffiti Data List.

6. **For Title, click Edit, and type** Graffiti Data for the Last 12 Weeks.

7. **Click the List tab, delete {ROUTE_NAME}, click the Insert Attribute button** {} ▾ , **select DATE_ CALL, and press Enter to get a new line in the rich text editor.**

8. **Click the Insert Attribute button, select ADDRESS, type a comma, press the space bar, click the Insert Attribute button again, and select ARTIST.**

9. **Click Done.**

10. **Click the Selector button for the new list, point to the Drag Item button** ✥ , **and drag the list to below the End Date list using Dock As Row.**

11. **Resize the two lists so that they appear as shown.** Do not choose Stack As for the new position. If you happen to stack the two lists so that you can view them by clicking tabs at the bottom of their column, drag the list again and reposition to unstack the two lists. Then you can view both at the same time.

Add actions to the graffiti data list

Next, you will configure the graffiti data list so that when you click a list item, the map pans and zooms to the location and flashes the point.

1. **Click the Selector button of the graffiti data list, and click Configure.**

2. **Click the Actions tab, click the Add Action drop-down list, select Pan, and for Add Target, select Graffiti.**

3. **Repeat step 2 two more times to add the actions Zoom and Flash.**

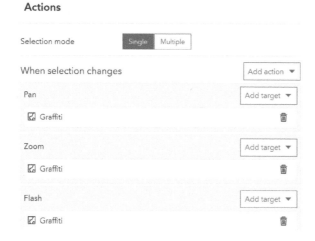

4. **Click Done.**

5. **Save your dashboard.**

> ## YOUR TURN
>
> In the list, click a record to see the map pan and zoom to the point, and flash. To return to the original extent, click the same point in the list again, or click the Pittsburgh bookmark in the upper-right menu of the map.

Add an action to the Artist Selector

Next, you will add an action to the Graffiti Artists query so that when you select an artist, only his or her graffiti displays in the graffiti data list.

1. **In the Graffiti Artists list, click the Selector button on the right of None, and click Configure.**

2. **Click Add Target and select Graffiti Data List.**

3. **Click Done.** Now when an artist is selected in the side panel, only his or her graffiti data will be listed (corresponding to the artist's mapped points).

4. **In the Graffiti Artists pane, select DKF.**

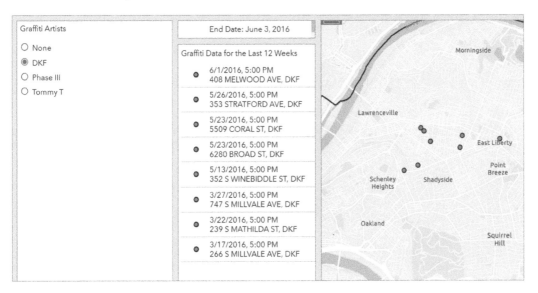

5. **In the Graffiti Artists pane, select None, and save your dashboard.**

Add a bar chart for graffiti type

1. **Click the Add Element button, select Serial Chart, and for layer, select Graffiti.**

2. **From the Data tab, in the Data Options panel, for Category Field, select GRAFFITI_TYPE_CODE.** Note that Count is the default statistic, which provides the desired frequency count for graffiti types in this case.

3. **For Sort By, select Statistic > Sort Descending, and for Maximum Categories, select 4.**

4. **Click the Category Axis tab, and for Title, type** Graffiti Type, **and for Placement, select Rotated.**

5. **Click the General tab, and for Title, type** Graffiti Type Bar Chart.

6. **For Name, type** Graffiti Type Bar Chart.

7. **Click Done.**

8. **Reposition the bar chart between the end date and data list using Dock As Row.**

9. **Resize the three elements as shown in the figure.**

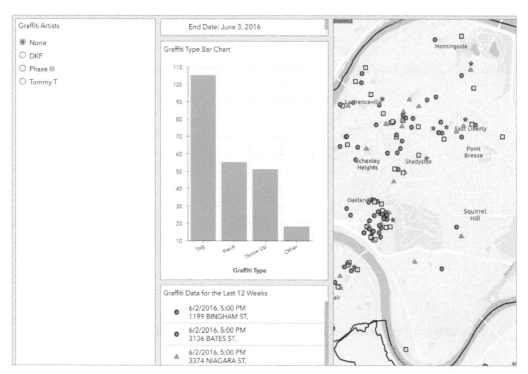

10. **Save your dashboard.**

YOUR TURN

Add Graffiti Type Bar Chart as a target to the Actions of the Artist Selector. Click Tommy T as the artist. The map and list show only graffiti for Tommy T, and the bar chart shows the corresponding bar chart for the mix of Tommy T's graffiti. When finished, select None for Artists to show results for all data.

Add an extent action to the map

The extent action for the map will limit the bar chart and graffiti data list to only those points in the current map extent. This configuration is useful for an investigator who wants to study the graffiti of a neighborhood.

1. Click the Selector button of the map, and click Configure.

2. Click the Map Actions horizontal menu item.

3. Click Add Action > Filter.

4. Click Add Target, and click Graffiti Type Bar Chart.

5. Repeat step 4, selecting Graffiti Data List.

6. Click Done. Now the bar chart and data list show data for only the current map extent.

7. Use the Central Business District bookmark. The bar chart and graffiti list show results for only the Central Business District map extent.

8. Zoom back to see all of Pittsburgh.

9. Save your dashboard.

Add a date selector

Often, investigators will want to study recently reported graffiti instead of the entire 12-week period available. For this purpose, you will add a date selector. The user can select a starting date for the displays from the available 12 weeks of data.

1. Click the Selector button of the Graffiti Dashboard heading, and click the Add Date Selector button.

2. **For Label, type** Beginning Date for Graffiti Data.

3. **For Type, select Date Picker.**

4. **For Operator, select Is Or Is After.**

5. **For Name, type** Beginning Date Selector.

6. **Click the Actions tab.**

7. **Click Add Action > Filter.**

8. **Click Add Target and select Graffiti ⊹ Graffiti, and for Filter Field, select DATE_CALL.**

9. **Similarly, add a target for Graffiti Data List with DATE_CALL as Filter Field, and add a target for Graffiti Type Bar Chart selecting DATE_CALL for Filter Field.**

10. **Click Done, save your dashboard, and try out the date selector for 5/24/2016. Make sure that None is selected for Graffiti Artists.**

11. **Scroll down the Graffiti Data List to see that the earliest CALL_DATE is 5/24/2016.**

YOUR TURN

Try changing map extent, graffiti artists, and beginning date to see that all work together as desired.

Use and share your dashboard

1. **Click the Home button** Home ▽, **and select Content.**

2. **Click Graffiti Dashboard Your Name.** Here, you have options to view the dashboard as any user would, edit the dashboard, and share it.

3. **Click Share, and if you are a member of an organization, you can select the option to share only with that organization or with everyone.** If you belong to any groups, you could share your dashboard with them.

    Note: To share your dashboard, you must also share the dashboard's web map and layers with your organization or groups.

4. **View your dashboard.** You can copy your dashboard's URL from your browser and send it to others so they can open and use it.

YOUR TURN

In edit mode, explore and use the options to format some of your dashboard elements. For example, for your header panel, change its size, text color, and background color.

Assignments

This chapter has two assignments to complete that you can download from ArcGIS Online, at go.esri.com/GISTforPro2.8Data:

- **Assignment 12-1:** Create tasks to publish a choropleth map for the Graffiti Dashboard operation view.
- **Assignment 12-2:** Modify and extend the Graffiti Dashboard operation view.

CHAPTER 13

Operations management with GIS
Graffiti Removal System

LEARNING GOALS

- Build ModelBuilder models to automate an operations management system.
- Use Network Analyst to optimize routes for carrying out service deliveries.
- Use Python expressions to calculate fields.
- Prepare data for use in the ArcGIS® Collector app.
- Prepare a map in ArcGIS Online for use in Collector.
- Use the Collector app to update data using a mobile device.

Introduction

This chapter is the second on GIS for operations management. Chapter 12 introduced operations management and GIS-based operations management systems for providing goods and services. In chapter 12, you built the Graffiti Mapping System, including tasks to prepare weekly data for mapping and an operations view using ArcGIS® Dashboards with a map, queries, and a bar chart useful for police in apprehending serial graffiti artists.

In this chapter, you'll build the Graffiti Removal System for a Public Works department supervisor, who oversees graffiti removal, and a Public Works employee, who removes the graffiti from public surfaces such as buildings, walls, and bridges. More specifically, in tutorials 13-1 through 13-3, you'll build the following three models that a Public Works department supervisor will run to schedule graffiti for removal at the beginning of each workweek:

- The **Identify Graffiti for Removal model** maps all sites that the supervisor has approved for graffiti removal.
- The **Calculate Optimal Route** model uses Network Analyst tools to calculate the optimal route for a set of graffiti sites the supervisor has selected for removal. The supervisor can create one or more routes for a week with multiple runs of this model. Each time a route is created, its graffiti sites are automatically removed from consideration in designing additional routes for the week.

- **The Record Route Results model** saves an optimal route as a feature class. Each time the supervisor calculates the route to follow, she uses this model to record its results, make a map of the route, and provide driving directions for the graffiti removal employee.

Note that the work is cumulative, so you must finish tutorial 13-1 to work on tutorial 13-2, and you must finish tutorial 13-2 to work on tutorial 13-3.

Before the actual graffiti removal, the supervisor must visit and survey each reported graffiti site. The supervisor may not approve graffiti removal for several reasons. For example, some 311 calls may give incorrect locations, some graffiti may have inaccessible locations, and some private property owners may not allow graffiti removal at their locations. In tutorial 13-4, you will build a Collector app so that the supervisor can use a mobile device, such as a smartphone, to enter data while inspecting a site. The supervisor classifies graffiti by type, determines a method for removal, estimates the work time, and enters other data. You can also go to ArcGIS Online to access assignments 13-1 and 13-2 to build models that create optimal routing for new graffiti site inspections and removal.

Tutorial 13-1: Build the Identify Graffiti for Removal model

This model resembles the model in tutorial 12-1 but includes new query criteria for sites that need graffiti removal.

Open the Tutorial 13-1 project

1. **Open Tutorial13-1.aprx from Chapter13, and save it as** Tutorial13-1YourName.aprx.

2. **Use the Pittsburgh bookmark.** The map has the following layers:
   - Garage, the location of the Public Works department garage that is the start and end of every route
   - GraffitiCalls, the up-to-date map layer of graffiti calls
   - Pittsburgh, the boundary of Pittsburgh
   - PittsburghStreets_ND, a network dataset based on Pittsburgh's TIGER streets

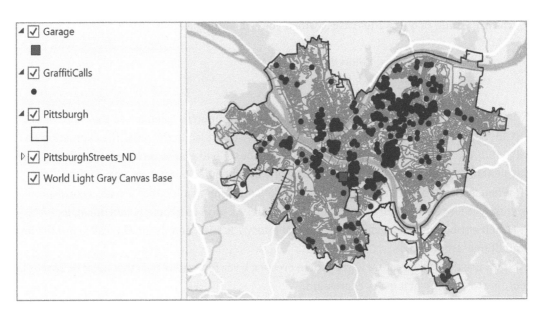

3. **In the Contents pane, turn off GraffitiCalls.**

Build a model to query sites for graffiti removal

The first model identifies two kinds of sites that need graffiti removed:

- Graffiti that the Public Works supervisor has approved but not yet scheduled for removal.
- Graffiti that has been scheduled for removal but not removed for some reason—for example, because an employee ran out of time or supplies.

Although it's possible to implement both criteria, for simplicity you'll implement only the first one, which is also the main one.

You will use three attributes of GraffitiCalls to retrieve graffiti for scheduled removal:

- Date_Approved is the date that a supervisor approved a 311 graffiti site to have its graffiti removed.
- Date_Scheduled is the date that the supervisor assigned to remove graffiti from a site.
- Date_Finished is the date that graffiti was removed from a site.

The expression of query 1 for case 1 is

```
DATE_APPROVED IS NOT NULL AND
DATE_SCHEDULED IS NULL
```

This compound query expression retrieves graffiti calls that have been approved but not yet scheduled for removal. If Date_Approved is not null, meaning that a date is entered, the graffiti is approved for removal by the supervisor. If Date_Scheduled is null, meaning that no date is entered for Date_Scheduled, the graffiti removal is not scheduled. With the expressions joined with the AND connector and when both expressions are true, the graffiti is a candidate for including in a route for removal.

The expression of query 2 for case 2, which you will not implement, is

```
DATE_SCHEDULED < CURRENT_DATE() AND
DATE_FINISHED IS NULL
```

CURRENT_DATE () has the current date from your computer's calendar, so the expression DATE_SCHEDULED < CURRENT_DATE() finds all graffiti sites scheduled in the past. The expression DATE_FINISHED IS NULL finds all graffiti sites that have not yet had graffiti removed. Together, connected with an AND, the expression also finds graffiti that was scheduled but not removed.

When queries 1 and 2 are combined with an OR, meaning that if either is true, a corresponding graffiti call is retrieved for scheduling a new route. As soon as a route is scheduled, its DATE_APPROVED attribute is given a date value, so that query 1 is false, and the graffiti call is not retrieved for additional scheduling.

You'll use the Make Feature Layer tool to create a layer for graffiti calls that must be scheduled according to only the first query expression.

1. **Open the Catalog pane, expand Toolboxes, right-click the Tutorial13.tbx toolbox, click New, and click Model.** The new, empty model opens in edit mode, ready for you to add model processes.

2. **In the Catalog pane, right-click Model, and click Properties. For Name, type** IdentifyGraffitiforRemoval, **and for Label, type** a. Identify Graffiti for Removal. **Click OK, and save your model.**

3. **On the Analysis tab, click Tools, search for the Make Feature Layer tool, drag it from the Geoprocessing pane to your model.**

4. **Click anywhere in the Model window to remove the selection handles from the Make Feature Layer process and its output.**

5. **Right-click the Make Feature Layer process in your model, and click Open.**

6. **For Input Features, select GraffitiCalls, and for Output Layer, type** GraffitiForRemoval. **Click New Expression, and create the query expression as shown in the figure.** Alternatively, you can click the SQL button, and copy and paste the query expression from chapter13\Tutorials\Resources\Tutorial13-1Query.txt.

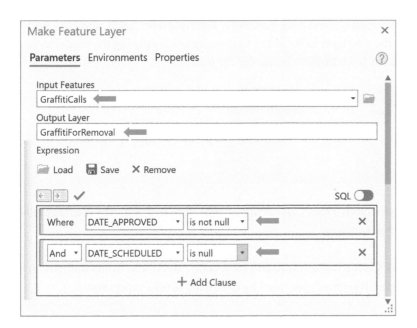

7.   **Click OK.** The model elements get color, indicating that they are ready to run.

8.   **Right-click the Make Feature Layer process, and click Run.** If your model has an error, the error will be printed in red text in the message window.

9.   **Close the Model message window.**

Symbolize output

1.   **Add the Apply Symbology From Layer tool to your model below the Make Feature Layer process.**

2.   **Drag an arrow from GraffitiForRemoval to Apply Symbology From Layer, and click Input Layer.**

3.   **Open Apply Symbology From Layer, and for the Symbology Layer input, browse to Chapter13\ Tutorials\Resources, and select Graffiti.lyrx.** The Graffiti.lyrx layer file has unique symbols for GRAFFITI_TYPE_CODE, plus labels for GraffitiForRemoval. The labels display work time estimates (minutes) for graffiti removal at each site.

4.   **Click OK and rearrange and resize elements of the model to improve its appearance.** Note that process inputs and outputs must have unique names in a model, even when they refer to the same feature class or map layer. It does not matter, for example, if your model has GraffitiForRemoval (3) instead of GraffitiForRemoval (2), as long as names are unique.

5. **Right-click GraffitiForRemoval (2), and click Parameter and right-click again and click Add To Display.** This will cause the new layer to be added to the map when opening the model to run it from the user interface that ModelBuilder builds for the model.

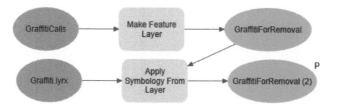

6. **Right-click the Apply Symbology From Layer process, click Run, and close the message window when the model is finished running.**

7. **On your map, right-click GraffitiForRemoval, and click Label.** The labels, from Graffiti.lyrx, show the estimated time for graffiti removal, provided by the Public Works supervisor.

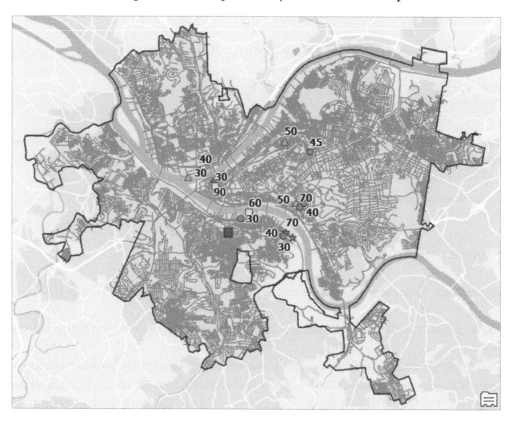

8. **On the ModelBuilder tab, click Save to save your model, and close it.** Note that if you need to edit your model again, right-click it in the Catalog pane, and click Edit.

Select graffiti sites for removal

Your map now has 14 sites with graffiti for removal. Next, the supervisor must review the map and decide on how many routes to create and which sites to include in each route. The graffiti removal employee has 480 minutes of work time per day. If the employee finishes a route early or there are days without routes, there is always other work for him to do. If the employee doesn't finish a route, the graffiti missed for removal will be scheduled in a future route. Scheduling within the 480-minute constraint takes some expertise and trial and error by the supervisor. After reviewing the map, suppose that she decides to make two routes for the 14 graffiti sites. The first route will have the eight sites indicated with arrows in the figure of step 2 that follows.

1. View your map, right-click GraffitiForRemoval, and click Zoom To Layer.

2. Using the Select tool under the Map tab, select the eight sites on the right side, as indicated in the figure.

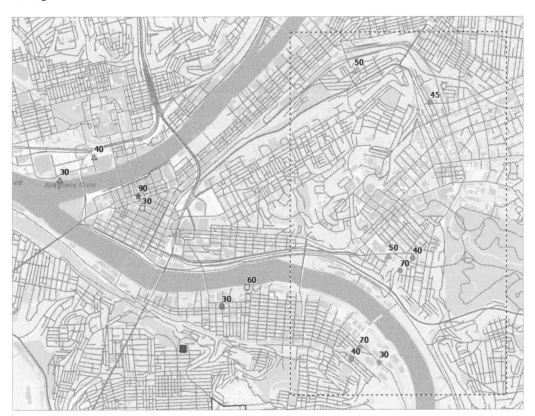

Tutorial 13-2: Build the Calculate Optimal Route model

This tutorial includes the standard workflow for calculating an optimal route (with minimum travel time) for a typical problem encountered in sales. In the prototypical problem, a salesperson leaves the office each day, has a list of customer addresses for stops, and returns to the office at the end of the route. Similarly, in this case, the graffiti truck leaves its garage, stops at an ordered list of graffiti sites along the optimal route, and returns to the garage.

Create a route

1. **Save your project as** Tutorial13-2YourName **in Chapter13\Tutorials.**

2. **In the Tutorial 13 toolbox, create a model named** CalculateOptimalRoute **and labeled**
    b. Calculate Optimal Route. **Click OK.**

3. **Add the Make Route Analysis Layer tool to the model, and open it.** This process, when run, creates a group layer in the Contents pane with standard map layers for the solution of a routing problem. The process's elements already have color fill, so you can run the process because it has default input street network data from ArcGIS.com. While working in this book, however, you will not use the default street network dataset because it consumes credits that require purchase. To learn more about credits, visit esri.com/en-us/arcgis/products/credits/overview. Next, you'll change the input network dataset to PittsburghStreets_ND, which is available and free in the data supplied with this book but less accurate than the default data source. Contact your instructor first if you want to use the default ArcGIS.com network dataset for a project or other work outside this book, unless you know you have Esri credits available for use.

4. **Make selections as shown for the Make Route Analysis Layer process.** The Preserve both first and last stops option creates a network problem for the prototypical traveling salesperson problem described at the beginning of this tutorial. The Accumulate Attributes option of Minutes causes route statistics, including total length, to be reported in minutes (instead of miles).
    *   **Network Data Source: PittsburghStreets_ND**
    *   **Sequence: Preserve both first and last stops**
    *   **Accumulate attributes: Minutes**

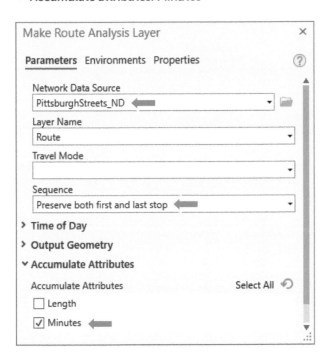

5. **Click OK.**

6. **Rearrange and resize inputs, temporarily move the PittsburghStreets_ND input up, delete Network Data Source, and rearrange the model elements as shown.**

Add locations

You must add the garage as the first location, then add the selected graffiti locations, and finally add the garage again as the last stop in the route.

1. **Add the Add Locations tool to your model three times, lining up the processes vertically.**

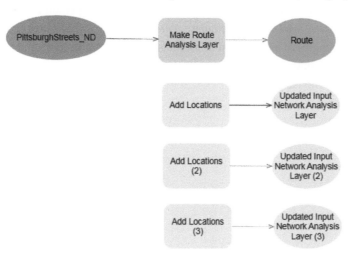

2. **Open the first Add Locations process. For Input Network Analysis Layer, select Route, and for Input Locations, select Garage.**

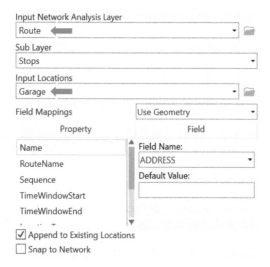

3. **Similarly, configure the Add Locations (2) process with Route (2) as Input Network Analysis Layer and GraffitiForRemoval as Input Locations.**

4. **Configure the Add Locations (3) process with Route (3) as Input Network Analysis Layer and Garage:1 as Input Locations.** As a reminder, every model element must have a unique name, hence the parentheses with digits after repeating names, such as Route (2) and Garage (2). It does not matter if the numbers in parentheses are in sequential order or if numbers are skipped, as long as all names are unique.

5. **Rearrange and resize model objects as shown.**

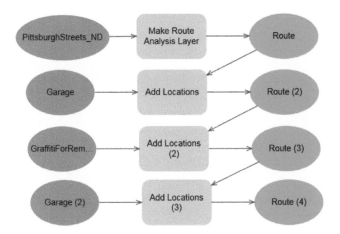

Add the Solve tool, and run the model

1. Add the Solve tool to your model, and open it.

2. For Input Network Analysis Layer, add Route (4), and click OK.

3. Right-click Route (5), and click Add To Display.

4. Right-click Route (5) again, and click Parameter.

5. Save your model.

6. On the ModelBuilder tab, in the Run group, click Run. Network Analyst found what appears to be a good route for the eight stops. If your route contains all the stops and not just the eight shown here, refer to the section "Select graffiti sites for removal," step 2, and rerun the model with only those sites selected. If the supervisor does not like this route for any reason, she can remove it from the Contents pane, select different sites from GraffitiForRemoval, and create a different route with the Create Route model. To rerun the model in edit mode, on the ModelBuilder tab, in the Run group, you'd first have to click Validate. Each time you create a route, a new route is saved to the Tutorial13.gdb file geodatabase with the alias name Route.

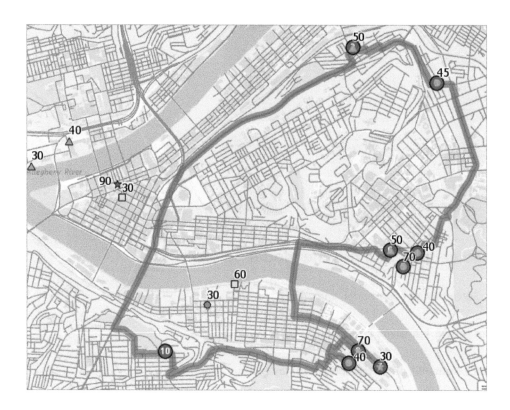

Tutorial 13-3: Build the Record Route Results model

Now that you have a route, you'll write driving directions for the graffiti removal employee, save the route with a name that includes the route's schedule date, and write a route name and schedule date in the master feature class, GraffitiCalls.

Create a model

1. **Save your project as** Tutorial13-3YourName **in Chapter13\Tutorials.**

2. **In the Tutorial13 toolbox, create a model named** RecordRouteResults **and labeled** c. Record Route Results.

3. **Add tools to your model as shown in the figure.** The Directions process will write driving instructions for the graffiti removal employee. The Copy Features process makes a permanent feature class for a route. The first Calculate Field process writes the new route's name in GraffitiCalls. The second Calculate Field process writes the scheduled date for the route in GraffitiCalls.

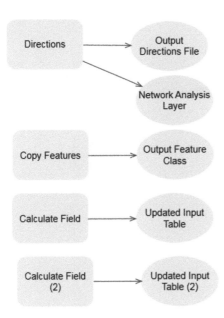

Create a variable

1. **On the ModelBuilder tab, in the Insert group, click Variable > OK to create a string data–type variable.** If you do not see the Insert group, you can widen your ArcGIS Pro window.

2. **Right-click the new variable, click Rename, and name the variable** RouteName.

3. **Open the variable, type** 20160613, **and click OK.** The date format, YYYYMMDD, has the desirable property that it sorts chronologically.

4. **Right-click the variable, and click Parameter.** When this model is open and run from the user interface created by ModelBuilder, the user can type a new date value.

Configure the Directions process

1. **Open Directions.**

2. **For Input Network Analysis Layer, select Route. For Output File Type, select Text, and for Output Directions File, type** Route%RouteName% **as shown.** The file name for the Output Directions file uses in-line substitution of the value stored in the RouteName variable for the text file name that is output from the Directions process. The percent delimiters around RouteName trigger the substitution, resulting in the file name Route20160613.txt.

Input Network Analysis Layer

Route (5):Route ⬅

Output File Type

Text ⬅

Output Directions File

C:\EsriPress\GISTforPro\Chapter13\Tutorials\Route%RouteName%.txt ⬅

Report Length in These Units

Miles

☑ Report Travel Time

Time Attribute

Minutes

Language

en

Style Name

Printable driving directions

Stylesheet

3. **Click OK.**

4. **Right-click and run the Directions process.**

5. **Open the output text file in Chapter 13\Tutorials to see the driving instructions, and close the file when you finish.** The document is a text document with the name Route20160613 and is not the XML document Route20160613.txt.

Configure the Copy Features process

1. **Open the Copy Features process.**

2. **For Input Features, select Route\Routes; for Output Feature Class, type** Route%RouteName%. The output feature class's name uses in-line substitution to include the date of the route.

3. **Click OK.**

4. **Run the Copy Features process.**

5. **From the Catalog pane, add the newly created feature class, Route20160613, to your map.** Note that the model may add the feature class to Contents with the label Route%RouteName%, instead of Route20160613. Nevertheless, the name of the stored feature class in Tutorial13.gdb is Route20160613.

6. **In the Contents pane in the Route group, turn off Routes.** Now you can see the newly created Route20160613.

Configure the first Calculate Field process

Recall that the source of the GraffitiForRemoval layer is the GraffitiCalls feature class. So when you change attribute values in GraffitiForRemoval, the changes are actually made in GraffitiCalls. GraffitiCalls is a master feature class that must be updated periodically with new 311 graffiti call data, at least weekly in this case. The updating work is not covered in this chapter, but you can do this work in a model using the Merge tool, in which update data is merged with the master feature class.

1. **Open the first Calculate Field process.**

2. **Type or make your selections as shown.** Ensure that you type the single quotation marks around Route%RouteName%. The Python computer language expression needs the quote marks to determine that Route%RouteName% is a string value.

   - **Input Table:** GraffitiForRemoval
   - **Field Name:** ROUTE_NAME
   - **ROUTE NAME** = 'Route%RouteName%'

Input Table

> GraffitiForRemoval (2):GraffitiForRemoval ◄▬▬

Field Name (Existing or New)

> ROUTE_NAME ◄▬▬

Expression Type

> Python 3

Expression

Fields ▼	Helpers ▼
OBJECTID	.as_integer_ratio()
Shape	.capitalize()
REQUEST_ID	.center()
ADDRESS	.conjugate()
ZIPCODE	.count()
DATE_TERMINATED	.decode()
WORK_TIME_ESTIMATE	.denominator()

Insert Values ▼   * / + - =

ROUTE_NAME =

> 'Route%RouteName%' ◄▬▬

3. **Click OK.**

4.   **Run the Calculate Field process.**

5.   **Open the GraffitiForRemoval attribute table and sort ROUTE_NAME descending to see that the process worked.** All eight selected graffiti sites have the route name, Route20160613.

6.   **Close the attribute table.**

Configure the second Calculate Field process

This process needs a Python computer language code block to reformat the string-value date, 20160613, to a date value, with the Python format 13/06/2016, for writing in the Date_Scheduled field. You'll copy and paste the code block (without having to type it). A version of the code with comments explaining syntax is available from Chapter13\Tutorials\Resources in the file Tutorial13-3CodeBlockComments.txt. The lines in that file starting with a hash sign are comments ignored by Python but available for you to read and understand the code. Note that you can learn some Python programming using free online tutorials from Esri (search Esri's training course catalog for Python: https://www.esri.com/training/catalog/search/).

1.   **Open the second Calculate Field process.**

2.   **For Input Table, select GraffitiForRemoval:1, and for Field Name, select DATE_SCHEDULED.**

3. **Open Chapter13\Tutorials\Resources\Tutorial13-3CodeBlock.txt, and copy and paste two sets of text as shown.** The `getdate` line after `DATE_SCHEDULED` runs the code block with the input value of RouteName in format YYYYMMDD. The variable `ds` in the code block returns a value for Date_Scheduled in the needed format of MM/DD/YYYY.

```
DATE_SCHEDULED = getdate('%RouteName%')
```

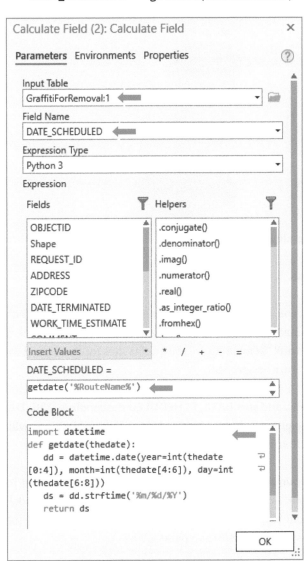

4. **Click OK.** You will see the completed model next.

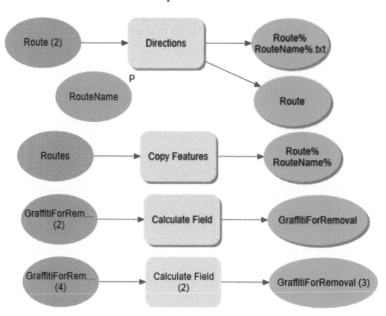

5. **Right-click GraffitiForRemoval (3), and click Add To Display.**

6. **Run the second Calculate Field process.**

7. **Open your map, and in the Contents pane, turn off the Route group.** You'll see that the graffiti sites in the route no longer display in GraffitiForRemoval, because those sites no longer meet the query criterion that Date_Scheduled is null. Remaining are the graffiti sites for the second route, which you'll schedule in the next Your Turn assignment.

8. **Open the attribute table for GraffitiCalls, and sort DATE_SCHEDULED descending.** With DATE_ SCHEDULED sorted, you can see the eight calls just scheduled for 6/13/2016. Next, you will configure model b. Calculate Optimal Route to run from its interface.

9. **In edit mode for model b. Calculate Optimal Route, right-click Route(5), and turn off Add To Display (Parameter is already turned on).** Add To Display is only useful when running models from edit mode. To add the route to your map while running the model from its interface (next in the Your Turn assignment), you must have Route(5) as a parameter.

10. **Save and close any open models.**

YOUR TURN

In your map, in the Contents pane, remove Route 20160613, and select the six remaining graffiti calls of the GraffitiForRemoval map layer. Then run your *b* and *c* models from the Catalog pane successively by right-clicking and clicking Open (and not Edit). When you run model *b*, there are no parameters for you to supply, so just click Run. When you open model *c*, type **20160614** for RouteName, and run the model. Look at your map and GraffitiCalls' attribute table for DATE_SCHEDULED and ROUTE_NAME, now 6/14/2016 and Route20160614, respectively. Save your project.

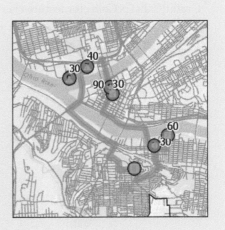

Tutorial 13-4: Update data using Collector

The Public Works supervisor, who oversees graffiti removal, must visit each graffiti site called in to 311 to make several assessments. She inspects graffiti and completes the following tasks, including entering data for the call in GraffitiCalls using Collector:

1. Correct the location of a graffiti call, if in error.
2. Decide whether to terminate the removal process (for example, because no graffiti exists), and if the call is terminated, entering the date in Date_Terminated.
3. Enter the date in Date_Approved if she plans to schedule the graffiti for removal.
4. Classify and select the graffiti type for Graffiti_Type_Code.
5. Enter an artist's tag or other identifier if the graffiti and its artist can be identified.
6. Choose the removal method, and select a value for Removal_Type_Code.
7. Estimate the time in minutes for removing the graffiti, and enter the value in Work_Time_Estimate.

You will create a map to use in Collector that displays only new graffiti calls that the supervisor must inspect and has a data entry form including the attributes she must enter. The form has drop-down lists for codes Graffiti_Type_Code and Removal_Type_Code to use in selecting code values instead of typing them.

As you will see, the knowledge and skills gained in this tutorial are for getting data ready for use in Collector.

Open the Tutorial 13-4 project

1. **Open Tutorial13-4 from Chapter13, and save it as** Tutorial13-4YourName.

2. **Use the Pittsburgh bookmark.** The GraffitiCalls map layer is made from the Tutorial13.gdb\
    GraffitiCallsForCollector feature class. This feature class does not have certain properties
    selected or set. Instead, you will learn to set the properties to enable capabilities in Collector.

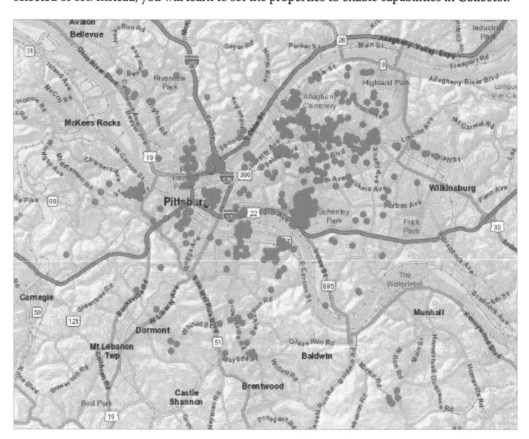

Create a coded value domain

For data input or editing in Collector and elsewhere, attributes of feature classes that are codes, such
as Removal_Type_Code and Graffiti_Type_Code, must have their values restricted to unique sets
of code values. For example, Removal_Type_Code must have values from only the set: Paint, Power
Wash, Sandblast, Scrape, and Solvent. Then when querying for graffiti calls, for example, that require
power washing for removal, you are certain that the criterion, `Removal_Type_Code = Power Wash`,
will retrieve all corresponding graffiti calls. There will be no calls with different spellings of *Power
Wash*, such as "Powerwash." Furthermore, data entry and editing are simplified, because instead of
typing, you select code values from a drop-down list in a data entry form. In ArcGIS Pro, unique code
values are implemented using coded value domains. Next, you'll create a coded value domain for
Removal_Type_Code.

1. **Right-click GraffitiCalls, and click Design > Domains.**

2. **Click the blank cell at the bottom of the Domain Name column, type** RemovalType, **and press Tab.** In step 3, you will refer to the upper-right portion of the next figure.

Domain Name	Description	Fie	Domain Type	Split Pc	Merg		Code	Description
NALocationStatus		Lon	Coded Value [	Default	Defau			
NALocationType		Lon	Coded Value [	Default	Defau			
NASideOfEdge		Lon	Coded Value [	Default	Defau			
PosAlong		Dou	Range Domair	Default	Defau			
Sequence		Lon	Range Domair	Default	Defau			
SourceID		Lon	Coded Value [	Default	Defau			
SourceOID		Lon	Range Domair	Default	Defau			
Total_Length		Dou	Range Domair	Geometi	Sum V			
Total_Minutes		Dou	Range Domair	Geometi	Sum V			
TotalViolation_Minutes		Dou	Range Domair	Geometi	Sum V			
TotalWait_Minutes		Dou	Range Domair	Geometi	Sum V			
Violation_Minutes		Dou	Range Domair	Geometi	Sum V			
Wait_Minutes		Dou	Range Domair	Geometi	Sum V			
RemovalType			Coded Value [	Default	Defau			

3. **In the pane on the right of the Domains table, for both Code and Description, type** Paint, **and press Enter.** You don't need a separate description for removal types because the coded values are self-descriptive. Nevertheless, ArcGIS Pro requires values entered for descriptions.

4. **Type four more codes as shown in the figure. When typing** Solvent **in the Description column, do not press Tab or Enter; instead, click Solvent in the Code column to enter the Solvent value in the Description column.** If you press Tab or Enter, a new code row opens, and you must type values in it, but Solvent is the last code value. If you accidentally add a new row after Solvent, you'll have to delete it.

Code	Description
Paint	Paint
Power Wash	Power Wash
Sandblast	Sandblast
Scrape	Scrape
Solvent	Solvent

5. **On the Domains tab, click Save.**

YOUR TURN

Create a second coded value domain, named **GraffitiType**. Add the following coded values, both for Code and Description: **Other**, **Piece**, **Tag**, and **Throw Up**. Save and close the Domains table when finished.

Apply a coded value domain

Now that you've defined domains, you must apply them to the Removal_Type_Code and Graffiti_Type_Code attributes in GraffitiCalls. If there was another feature class that had either of the codes as an attribute (there isn't in this case), you'd apply the corresponding domain to it as well.

1. In the Contents pane, right-click GraffitiCalls, and click Design > Fields.

2. Click the empty cell at the intersection of the REMOVAL_TYPE_CODE row and Domain column, and select RemovalType. You may have to click a few times before the drop list of available domains appears.

3. On the Fields tab, click Save.

YOUR TURN

Apply the GraffitiType domain to GRAFFITI_TYPE_CODE in the GraffitiCalls feature class. Save and close the Fields view.

Rename a feature class

If you are in a class with students all publishing to the same ArcGIS Online organizational account, each feature class published must have a unique name for all students. You'll publish the GraffitiCalls feature class in the exercise after this one, but first you must give it a unique name.

1. Open the Catalog pane.

2. Expand Databases > Tutorial13.gdb.

3. Rename GraffitiCallsForCollector GraffitiCallsYourName, substituting your name or ID. If more than one person has the same first name in your class, you can add a middle name or otherwise make your name unique.

4. Add GraffitiCallsYourName to the Contents pane, and remove GraffitiCalls.

5. Symbolize GraffitiCallsYourName to have a Circle 1 point symbol and a dark-blue color and size 5.

Publish a web layer

Next, you'll publish GraffitiCallsYourName to your ArcGIS Online organizational account. Then, after a few additional steps, it will be available for use in Collector by the Public Works supervisor in charge of graffiti removal.

1.   **In the Contents pane, right-click GraffitiCallsYourName, and click Sharing > Share As Web Layer.**

2.   **In the Share As Web Layer pane, for Summary, type** Up-to-date graffiti calls, **and for Tags, type**
     Graffiti **and** Pittsburgh. **Under Share With, check the box for Everyone.**

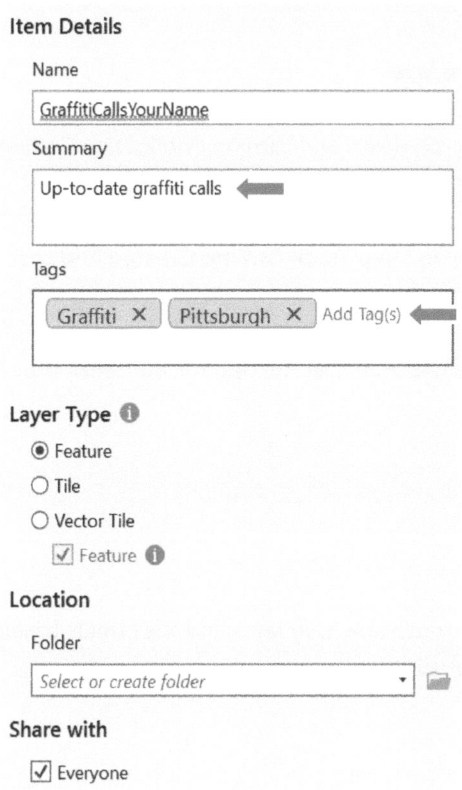

Item Details

Name

GraffitiCallsYourName

Summary

Up-to-date graffiti calls ⬅

Tags

Graffiti ✕   Pittsburgh ✕   Add Tag(s) ⬅

Layer Type ⓘ

⦿ Feature

◯ Tile

◯ Vector Tile

☑ Feature ⓘ

Location

Folder

Select or create folder ▾ 📁

Share with

☑ Everyone

3.   **Click Publish.**

4.   **After you publish the web layer, save and close your project and ArcGIS Pro.**

Set the editing option for a web layer in ArcGIS Online

First, you must allow the supervisor to edit the layer so that she can enter data into the GraffitiCalls web layer.

1. Sign in to your organizational account in ArcGIS Online.

2. In Content, click the GraffitiCallsYourName feature layer.

3. In the upper right of the window, click Settings, scroll down and turn on Public Data Collection, and then click Save.

4. Scroll down and turn on Enable Editing. Then turn on Keep Track Of Who Created And Last Updated Features.

5. Under What Kind Of Editing Is Allowed, make sure that Add, Delete, Update, and Attributes And Geometry are selected.

6. Click Save.

Create a map in ArcGIS Online

For this simple map, you will have just the GraffitiCallsYourName web layer and the Streets basemap, but you will add a filter to select only calls that need the supervisor's review.

1. Click the Overview tab.

2. Open your map in Map Viewer.

3. **Click the Basemap button, and change the basemap to Streets.**

4. **Click the Save button, type the text as shown, and click Save Map.**
    - **Title:** Graffiti Calls for Supervisor Mary Smith
    - **Tags:** Graffiti, Pittsburgh
    - **Summary:** Graffiti calls to be reviewed by supervisor.

Title

Graffiti Calls for Supervisor ⬅

Tags

Graffiti ✕ Pittsburgh ✕ *Add tag(s)* ⬅

Summary

Graffiti Calls to be reviewed by supervisor ⬅

Add a filter

For the filter (query), you'll select records for Date_Approved and Date_Terminated that are Null (blank). The selected records will correspond to new graffiti calls that have not been reviewed by the supervisor and have not been deemed invalid or duplicate calls.

1. **Click the Layers button, and select the GraffitiCallsYourName layer.**

2. **On the menu on the right, click Filter > Add Expression.**

3. **For DATE_APPROVED, select is blank.**

DATE_APPROVED

is blank ▼

4. **Click Add Expression, and for DATE_TERMINATED, select is blank.**

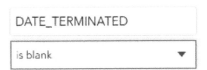

DATE_TERMINATED

is blank ▼

5. **Click Save.** You will see seven graffiti calls for the supervisor to review.

Resymbolize GraffitiCalls

The layer needs a larger symbol for viewing.

1. On the menu on the right, click Styles > Style Options.

2. Click the current symbol.

3. Under Symbols, choose the circular symbol, add a check mark to adjust size automatically, and red for color.

4. Click Done, and click Done to close the style menu.

5. Save the map.

Configure pop-up windows

The data entry form for Collector is based on the pop-up window that you configure next for GraffitiCallsYourName. The attributes of the layer that you will choose for the pop-up window are the ones that will appear on Collector's form, which Collector automatically generates from the pop-up.

1. On the menu on the right, click Configure Pop-ups.

2. Click Fields List > Select Fields.

3. Modify to display only the following attributes: ADDRESS, ZIPCODE, DATE_TERMINATED, WORK_TIME_ESTIMATE, COMMENT, DATE_CALL, DATE_APPROVED, REMOVAL_TYPE_CODE,

GRAFFITI_TYPE_CODE, and ARTIST. The supervisor can see and enter or modify these attributes on a form in Collector.

4. **Click Done.**

5. **Using the six dots on the left of each attribute, drag attributes to match the following order: ADDRESS, ZIPCODE, DATE_CALL, DATE_APPROVED, GRAFFITI_TYPE_CODE, REMOVAL_TYPE_ CODE, ARTIST, WORK_TIME_ESTIMATE, DATE_TERMINATED, and COMMENT.**

6. **Change the title to** Graffiti Sites for Review, **and click OK.**

7. **Save your map.**

Check map setting for use in Collector

1. On the upper-left menu, click the three bars and select Content.

2. Click Graffiti Calls for Supervisor.

3. In top right of window, click Settings.

4. Scroll down and make sure that Use In ArcGIS Collector is checked. Your map is ready for use in Collector.

5. Sign out of ArcGIS Online.

Get started with Collector

Collector is a free app available for your Windows 10, Android, or Apple iOS device (on the Windows Store, Google Play, or App Store, respectively).

1. Install Collector on your device.

2. Open the app.

3. Sign in using your ArcGIS Online user name and password.

4. In Maps, tap Graffiti Calls for Supervisor to open it, don't allow Collector to access your location, and zoom and pan to see the seven graffiti calls. Your map opens on your device, displaying the seven graffiti sites for review.

Use the Collector app to edit data

Next are general instructions for editing an existing graffiti point's location and data record. Whatever mobile device you have, do the following:

1. Tap the northernmost graffiti point to select it and make its pop-up visible. It's useful that both the map and attributes are visible, and that you can move the top of the form up or down.

2. Click the Edit button to start editing the graffiti location or attribute values. The point gets a target around it, and you can drag to change the location of the point.

3. Move the graffiti point to a new distinctive location that you can remember, and tap Update Point.

4.   **Select Tag for GRAFFITI_TYPE_CODE.**

5.   **Select Sandblast for REMOVAL_TYPE_CODE.**

6.   **Enter** 30 **for WORK_TIME_ESTIMATE.** Normally the supervisor would select date values for DATE_APPROVED or DATE_TERMINATED, but don't do that, because then the point would no longer be visible in your Graffiti Calls for Supervisor Your Name map in ArcGIS Online. You set a filter previously to show only graffiti points that have no values for those two dates.

7.   **Submit the edits.** Collector saves the edits in ArcGIS Online for you.

8.   **Open ArcGIS Online, and view your Graffiti Calls for Supervisor Your Name map.** See that the point you edited is in its new location. Also use the pop-up for the point to see that it has your edited attribute values.

Use a hosted feature layer for ArcGIS Pro processing

One step remains—namely, to integrate GIS processing in ArcGIS Pro and ArcGIS Online for GraffitiCalls. In a real operations management system, some GIS processing of GraffitiCalls would be done using both desktop and online/mobile systems. For example, new graffiti calls may be merged with GraffitiCalls using ArcGIS Pro and also scheduled, as done in tutorials 13-1 through 13-3. Then the supervisor and the employee who removes the graffiti would update GraffitiCalls records using Collector and ArcGIS Online. You will integrate the separate processing of desktop and online systems by using the hosted online feature layer as the source for GraffitiCalls in ArcGIS Pro.

1.   **Open Tutorial13-4YourName in ArcGIS Pro.**

2.   **Open Properties for GraffitiCallsYourName, and view its source.** The current version of GraffitiCallsYourName resides on your computer's hard drive, whereas a separate updated version resides in ArcGIS Online.

3.   **Close the Properties pane.**

4.   **Remove GraffitiCallsYourName from the Contents pane.**

5.   **On the Map tab, click the Add Data button.**

6.   **In Portal, click My Content, click GraffitiCallsYourName, and click OK.** Now ArcGIS Pro has the version of GraffitiCallsYourName that was updated by Collector. You can add or modify the same GraffitiCallsYourName feature layer using ArcGIS Pro.

YOUR TURN

In ArcGIS Pro, open the GraffitiCallsYourName attribute table. Find some changes that were made using Collector. Make a change of one cell's value in a GraffitiCallsYourName record, note the changed record, and close the table. Open your ArcGIS Online account, go to My Content, open the item page of the GraffitiCallsYourName feature layer, and click the Data tab. Scroll down in the table, if necessary, to the record you changed to see the change.

Assignments

This chapter has two assignments to complete that you can download from ArcGIS Online, at go.esri.com/GISTforPro2.8Data:

- **Assignment 13-1**: Build models for routing the supervisor's assessment of new graffiti.
- **Assignment 13-2**: Create a Collector map for the Public Works employee who removes graffiti.

Data source credits

Chapter 1
\Tutorials\Chapter1.gdb\AllCoTracts, courtesy of US Census Bureau.

\Tutorials\Chapter1.gdb\AllCoTracts_Statistics, courtesy of W. L. Gorr, Carnegie Mellon University.

\Tutorials\Chapter1.gdb\AlleghenyCounty, courtesy of US Census Bureau.

\Tutorials\Chapter1.gdb\FQHC, courtesy of W. L. Gorr, Carnegie Mellon University.

\Tutorials\Chapter1.gdb\FQHCBuffer, courtesy of W. L. Gorr, Carnegie Mellon University.

\Tutorials\Chapter1.gdb\MedExpress, courtesy of W. L. Gorr, Carnegie Mellon University.

\Tutorials\Chapter1.gdb\MedExpressBuffer, courtesy of W. L. Gorr, Carnegie Mellon University.

\Tutorials\Chapter1.gdb\Municipalities, courtesy of US Census Bureau.

\Tutorials\Chapter1.gdb\Parks, courtesy of Southwestern Pennsylvania Commission.

\Tutorials\Chapter1.gdb\Pittsburgh, courtesy of US Census Bureau.

\Tutorials\Chapter1.gdb\PovertyData, courtesy of US Census Bureau.

\Tutorials\Chapter1.gdb\PovertyIndex, courtesy of W. L. Gorr, Carnegie Mellon University.

\Tutorials\Chapter1.gdb\PovertyRiskContour, courtesy of W. L. Gorr, Carnegie Mellon University.

\Tutorials\Chapter1.gdb\Rivers, courtesy of US Census Bureau.

\Tutorials\Chapter1.gdb\Streets, courtesy of US Census Bureau, TIGER.

\Tutorials\Chapter1.gdb\XYAllCoTractsCentroids, courtesy of US Census Bureau and W. L. Gorr, Carnegie Mellon University.

Chapter 2
\Tutorials\Chapter2.gdb\Boroughs, courtesy of Department of City Planning, New York City.

\Tutorials\Chapter2.gdb\Facilities, courtesy of Department of City Planning, New York City.

\Tutorials\Chapter2.gdb\FireCompanies, courtesy of Department of City Planning, New York City.

\Tutorials\Chapter2.gdb\FireHouses, courtesy of Department of City Planning, New York City.

\Tutorials\Chapter2.gdb\ManhattanStreets, courtesy of Department of City Planning, New York City.

\Tutorials\Chapter2.gdb\Neighborhoods, courtesy of Department of City Planning, New York City.

\Tutorials\Chapter2.gdb\PolicePrecincts, courtesy of Department of City Planning, New York City.

\Tutorials\Chapter2.gdb\PoliceStations, courtesy of Department of City Planning, New York City.

\Tutorials\Chapter2.gdb\Water, courtesy of Department of City Planning, New York City.

\Tutorials\Chapter2.gdb\ZoningLandUse, courtesy of Department of City Planning, New York City.

Chapter 3
\Tutorials\Chapter3.gdb\MetroArtsPoints, courtesy of Bureau of Labor Statistics.

\Tutorials\Chapter3.gdb\MetroCOLIPoints, courtesy of The Council for Community and Economic Research.

\Tutorials\Chapter3.gdb\Neighborhoods, courtesy of The City of Pittsburgh, Department of City Planning.

\Tutorials\Chapter3.gdb\PghStreets, courtesy of US Census Bureau.

\Tutorials\Chapter3.gdb\Pittsburgh, courtesy of US Census Bureau.

\Tutorials\Chapter3.gdb\Rivers, courtesy of US Census Bureau.

\Tutorials\Chapter3.gdb\USStates, courtesy of US Census Bureau, The Council for Community and Economic Research, Bureau of Labor Statistics.

\Tutorials\Chapter3.gdb\USStatesPoints, courtesy of US Census Bureau, The Council for Community and Economic Research, Bureau of Labor Statistics.

Chapter 4
\Data\AlleghenyCounty\EducationalAttainment.csv, courtesy of US Census Bureau.

\Data\AlleghenyCounty\tl_2010_42003-tract10.shp, courtesy of US Census Bureau.

\Data\MaricopaCounty\CensusData.csv, courtesy of US Census Bureau.

\Data\MaricopaCounty\tl_2010_04013_cousub10.shp, courtesy of US Census Bureau.

\Data\MaricopaCounty\tl_2010_04013_tract10.shp, courtesy of US Census Bureau.

\Data\Pittsburgh\City.gdb\CrimeOffenses, courtesy of City of Pittsburgh Police Bureau.

\Data\Pittsburgh\City.gdb\Neighborhoods, courtesy of The City of Pittsburgh, Department of City Planning.

\Data\Pittsburgh\City.gdb\PghStreets, courtesy of US Census Bureau.

\Data\Pittsburgh\City.gdb\PghTracts, courtesy of US Census Bureau.

\Data\Pittsburgh\City.gdb\Pittsburgh, courtesy of US Census Bureau.

\Data\Pittsburgh\PittsburghSeriousCrimesSummer2015.shp, courtesy of City of Pittsburgh Police Bureau.

\Data\Pittsburgh\PovertyTracts.csv, courtesy of US Census Bureau.

\Data\Crime.gdb\Burglaries, courtesy of City of Pittsburgh Police Bureau.

\Data\Crime.gdb\CrimeOffenses, courtesy of City of Pittsburgh Police Bureau.

\Data\Crime.gdb\Neighborhoods, courtesy of The City of Pittsburgh, Department of City Planning.

\Data\Crime.gdb\Streets, courtesy of US Census Bureau.

Chapter 5
\Data\NewYorkCity\CouncilDistricts.shp, courtesy of Department of City Planning, New York City.

\Data\NewYorkCity\Libraries.dbf, courtesy of Department of City Planning, New York City.

\Tutorials\Chapter5.gdb\Counties, from ArcGIS® Data and Maps (2010), courtesy of ArcUSA, US Census Bureau.

\Tutorials\Chapter5.gdb\Country, from ArcGIS Data and Maps (2004), courtesy of *ArcWorld Supplement*.

\Tutorials\Chapter5.gdb\HennepinCounty, courtesy of US Census Bureau.

\Tutorials\Chapter5.gdb\Municipalities, courtesy of Southwestern Pennsylvania Commission.

\Tutorials\Chapter5.gdb\Ocean, from ArcGIS Data and Maps, courtesy of Esri.

\Tutorials\Chapter5.gdb\Parks, courtesy of Southwestern Pennsylvania Commission.

\Tutorials\Chapter5.gdb\Tracts, courtesy of US Census Bureau.

\Tutorials\Chapter5.gdb\States, from ArcGIS Data and Maps (2010), courtesy of ArcUSA, US Census Bureau.

Chapter 6

\Tutorials\Chapter6.gdb\Boroughs, courtesy of Department of City Planning, New York City.

\Tutorials\Chapter6.gdb\BronxWater, courtesy of US Census Bureau.

\Tutorials\Chapter6.gdb\BronxWaterfrontParks, courtesy of Department of City Planning, New York City.

\Tutorials\Chapter6.gdb\BrooklynWater, courtesy of US Census Bureau.

\Tutorials\Chapter6.gdb\BrooklynWaterfrontParks, courtesy of Department of City Planning, New York City.

\Tutorials\Chapter6.gdb\EMSFacilities, courtesy of Department of City Planning, New York City.

\Tutorials\Chapter6.gdb\FireCompanies, courtesy of Department of City Planning, New York City.

\Tutorials\Chapter6.gdb\FireHouses, courtesy of Department of City Planning, New York City.

\Tutorials\Chapter6.gdb\ManhattanBlockGroups, courtesy of US Census Bureau.

\Tutorials\Chapter6.gdb\ManhattanFireCompanies, courtesy of Department of City Planning, New York City.

\Tutorials\Chapter6.gdb\ManhattanLandUse, courtesy of Department of City Planning, New York City.

\Tutorials\Chapter6.gdb\ManhattanStreets, courtesy of US Census Bureau.

\Tutorials\Chapter6.gdb\ManhattanTracts, courtesy of US Census Bureau.

\Tutorials\Chapter6.gdb\ManhattanWater, courtesy of US Census Bureau.

\Tutorials\Chapter6.gdb\ManhattanWaterfrontParks, courtesy of Department of City Planning, New York City.

\Tutorials\Chapter6.gdb\NeighborhoodsZoningLandUse, courtesy of Department of City Planning, New York City.

\Tutorials\Chapter6.gdb\NYCBlockGroups, courtesy of US Census Bureau.

\Tutorials\Chapter6.gdb\NYCNeighborhoods, courtesy of US Census Bureau.

\Tutorials\Chapter6.gdb\NYCWaterfrontParks, courtesy of Department of City Planning, New York City.

\Tutorials\Chapter6.gdb\PoliceStations, courtesy of Department of City Planning, New York City.

\Tutorials\Chapter6.gdb\QueensWater, courtesy of US Census Bureau.

\Tutorials\Chapter6.gdb\QueensWaterfrontParks, courtesy of Department of City Planning, New York City.

\Tutorials\Chapter6.gdb\StatenIslandWater, courtesy of US Census Bureau.

\Tutorials\Chapter6.gdb\StatenIslandWaterfrontParks, courtesy of Department of City Planning, New York City.

\Tutorials\Chapter6.gdb\UpperWestSideFireCompanies, courtesy of Department of City Planning, New York City.

\Tutorials\Chapter6.gdb\UpperWestSideTracts, courtesy of US Census Bureau.

\Tutorials\Chapter6.gdb\UpperWestSideZoningLandUse, courtesy of Department of City Planning, New York City.

\Tutorials\Chapter6.gdb\ZoningLandUse, courtesy of Department of City Planning, New York City.

Chapter 7

\Data\HBH1.dwg, courtesy of Carnegie Mellon University.

\Tutorials\Chapter7.gdb\Bldgs, courtesy of The City of Pittsburgh, Department of City Planning.

\Tutorials\Chapter7.gdb\BldgsOriginal, courtesy of The City of Pittsburgh, Department of City Planning.

\Tutorials\Chapter7.gdb\BusStopCrossWalk, courtesy of Kristen Kurland, Carnegie Mellon University.

\Tutorials\Chapter7.gdb\Greenspaces (from Parks), courtesy of Southwestern Pennsylvania Commission.

\Tutorials\Chapter7.gdb\Streets, courtesy of US Census Bureau.

\Tutorials\Chapter7.gdb\StudyAreaBldgs, courtesy of The City of Pittsburgh, Department of City Planning.

\Tutorials\Chapter7.gdb\Water, courtesy of US Census Bureau.

Chapter 8

\Data\AliasTable.csv, courtesy of W. L. Gorr, Carnegie Mellon University.

\Data\AssignmentsData.gdb\PghStreets, courtesy of US Census Bureau, TIGER.

\Data\AssignmentsData.gdb\PghTracts, courtesy of US Census Bureau, TIGER.

\Data\AssignmentsData.gdb\Pittsburgh, courtesy of US Census Bureau, TIGER.

\Data\AttendeesAlleghenyCounty.csv, courtesy of FLUX.

\Data\AttendeesAlleghenyPARegion.csv, courtesy of FLUX.

\Data\Clients.csv, courtesy of Kristen Kurland, Carnegie Mellon University.

\Data\GroceryStores.csv, courtesy of W. L. Gorr, Carnegie Mellon University.

\Tutorials\Chapter8.gdb\AlleghenyCounty.AllCoZIP, courtesy of US Census Bureau, TIGER.

\Tutorials\Chapter8.gdb\AlleghenyCounty.AlleghenyCounty, courtesy of US Census Bureau, TIGER.

\Tutorials\Chapter8.gdb\AlleghenyCounty.Municipalities, courtesy of US Census Bureau, TIGER.

\Tutorials\Chapter8.gdb\AlleghenyCounty.Streets, courtesy of US Census Bureau, TIGER.

\Tutorials\Chapter8.gdb\PARegion.PARegion, courtesy of US Census Bureau, TIGER.

\Tutorials\Chapter8.gdb\PARegion.PARegionZIP, courtesy of US Census Bureau, TIGER.

\Tutorials\Chapter8.gdb\PittsburghCBD.CBDOutline, courtesy of The City of Pittsburgh, Department of City
     Planning.

\Tutorials\Chapter8.gdb\PittsburghCBD.CBDStreets, courtesy of The City of Pittsburgh, Department of City
     Planning.

\Tutorials\Chapter8.gdb\USStates, courtesy of US Census Bureau, TIGER.

Chapter 9

\Data\Exponential.xlsx, courtesy of W. L. Gorr, Carnegie Mellon University.

\Data\PittsburghNetworkDataset.gdb\PittsburghStreets\PittsburghStreets, courtesy of US Census Bureau,
     TIGER.

\Data\PittsburghNetworkDataset.gdb\PittsburghStreets\PittsburghStreets_ND, courtesy of US Census Bureau,
     TIGER, and W. L. Gorr, Carnegie Mellon University.

\Data\PittsburghNetworkDataset.gdb\PittsburghStreets\PittsburghStreets_ND_Junctions, courtesy of US
     Census Bureau, TIGER, and W. L. Gorr, Carnegie Mellon University.

\Tutorials\Chapter9.gdb\DrugViolations, courtesy of City of Pittsburgh Police Bureau.

\Tutorials\Chapter9.gdb\Neighborhoods, courtesy of The City of Pittsburgh, Department of City Planning.

\Tutorials\Chapter9.gdb\Pittsburgh, courtesy of US Census Bureau, TIGER.

\Tutorials\Chapter9.gdb\PittsburghBlockCentroids, courtesy of US Census Bureau, TIGER.

\Tutorials\Chapter9.gdb\PoliceZones, courtesy of The City of Pittsburgh, Department of City Planning.

\Tutorials\Chapter9.gdb\Pools, courtesy of Pittsburgh CitiParks Department.

\Tutorials\Chapter9.gdb\Pooltags, courtesy of Pittsburgh CitiParks Department.

\Tutorials\Chapter9.gdb\PovertyRiskContour, courtesy of US Census Bureau, TIGER, and W. L. Gorr, Carnegie
     Mellon University.

\Tutorials\Chapter9.gdb\Rivers, courtesy of US Census Bureau, TIGER.

\Tutorials\Chapter9.gdb\Schools, courtesy of The City of Pittsburgh, Department of City Planning.

\Tutorials\Chapter9.gdb\SeriousViolentCrimes, courtesy of City of Pittsburgh Police Bureau.

Chapter 10

\Data\LandUse_Pgh.tif, image courtesy of US Geological Survey, Department of the Interior/USGS.

\Data\LandUse.lyr, courtesy of US Geological Survey, Department of the Interior/USGS, and W. L. Gorr,

\Data\LandUse_Pgh.prj, courtesy of W. L. Gorr, Carnegie Mellon University.

\Tutorials\Chapter10.gdb\AllCoBlkGrps, courtesy of US Census Bureau, TIGER.

\Tutorials\Chapter10.gdb\Municipalities, courtesy of US Census Bureau, TIGER.

\Tutorials\Chapter10.gdb\NED, courtesy of US Geological Survey, Department of the Interior/USGS.

\Tutorials\Chapter10.gdb\Neighborhoods, courtesy of The City of Pittsburgh, Department of City Planning.

\Tutorials\Chapter10.gdb\OHCA, courtesy of Children's Hospital of Pittsburgh.

\Tutorials\Chapter10.gdb\Pittsburgh, courtesy of US Census Bureau, TIGER.

\Tutorials\Chapter10.gdb\PittsburghBlkGrps, courtesy of US Census Bureau, TIGER.

\Tutorials\Chapter10.gdb\PittsburghBlks, courtesy of US Census Bureau, TIGER.

\Tutorials\Chapter10.gdb\Rivers, courtesy of US Census Bureau, TIGER.

\Tutorials\Chapter10.gdb\ZoningCommercialBuffer, courtesy of The City of Pittsburgh, Department of City Planning, and W. L. Gorr, Carnegie Mellon University.

Chapter 11

\Data\LASFiles\1336704E409152N, courtesy of Pictometry International Corp.

\Data\LASFiles\1336704E411792N, courtesy of Pictometry International Corp.

\Data\LASFiles\1339344E409152N, courtesy of Pictometry International Corp.

\Data\LASFiles\1339344E411792N, courtesy of Pictometry International Corp.

\Data\LASFiles\1341984E409152N, courtesy of Pictometry International Corp.

\Data\LASFiles\1341984E411792N, courtesy of Pictometry International Corp.

\Data\PACPIT14_LiDAR_Delivery_Area.shp, courtesy of Pictometry International Corp.

\Tutorials\Chapter11.gdb\Bldgs, courtesy of The City of Pittsburgh, Department of City Planning.

\Tutorials\Chapter11.gdb\Contours, courtesy of The City of Pittsburgh, Department of City Planning.

\Tutorials\Chapter11.gdb\Courthouse3D, courtesy of The City of Pittsburgh, Department of City Planning.

\Tutorials\Chapter11.gdb\Courthouse3DTowers, courtesy of Southwestern Pennsylvania Commission.

\Tutorials\Chapter11.gdb\Parks, courtesy of Southwestern Pennsylvania Commission.

\Tutorials\Chapter11.gdb\ParkTrees, courtesy of The City of Pittsburgh, Department of City Planning.

\Tutorials\Chapter11.gdb\Rivers, courtesy of Southwestern Pennsylvania Commission.

\Tutorials\Chapter11.gdb\SmithfieldBldgs, courtesy of The City of Pittsburgh, Department of City Planning.

\Tutorials\Chapter11.gdb\SmithfieldTextured1, courtesy of Kristen Kurland, Carnegie Mellon University.

\Tutorials\Chapter11.gdb\SmithfieldTextured2, courtesy of Kristen Kurland, Carnegie Mellon University.

\Tutorials\Chapter11.gdb\StreetFurniture_Smithfield, courtesy of Kristen Kurland, Carnegie Mellon University.

\Tutorials\Chapter11.gdb\StreetCurbs, courtesy of The City of Pittsburgh, Department of City Planning.

\Tutorials\Chapter11.gdb\StreetTrees, courtesy of The City of Pittsburgh, Department of City Planning.

\Tutorials\Chapter11.gdb\StudyArea (selected features from PACPIT14_LiDAR_Delivery_Area), courtesy of Pictometry International Corp.

\Tutorials\Chapter11.gdb\USSteelBldg, courtesy of The City of Pittsburgh, Department of City Planning.

Chapter 12

\Data\Graffiti20160603.txt, courtesy of The City of Pittsburgh, Department of Innovation and Performance, downloaded from the Western Pennsylvania Data Center.

\Data\Graffiti20160610.txt, courtesy of The City of Pittsburgh, Department of Innovation and Performance, downloaded from the Western Pennsylvania Data Center.

\Tutorials\Chapter12.gdb\Graffiti, courtesy of The City of Pittsburgh, Department of Innovation and Performance, downloaded from the Western Pennsylvania Data Center, with Creative Commons license at https://data.wprdc.org/dataset/311-data. W. L. Gorr assigned random point locations within correct neighborhoods to the neighborhood-level data, added attributes (GraffitiTypeCode, RemovalTypeCode, Artist, Comment, and several dates used for scheduling graffiti removal in fictional scenarios), and entered fictitious values.

\Tutorials\Chapter12.gdb\Heading, courtesy of W. L. Gorr.

\Tutorials\Chapter12.gdb\Pittsburgh, courtesy of The City of Pittsburgh, Department of City Planning.

Chapter 13

\Data\Graffiti20160603.txt, courtesy of The City of Pittsburgh, Department of Innovation and Performance, downloaded from the Western Pennsylvania Data Center.

\Data\Graffiti20160610.txt, courtesy of The City of Pittsburgh, Department of Innovation and Performance, downloaded from the Western Pennsylvania Data Center.

\Tutorials\Chapter13.gdb\PittsburghStreets\PittsburghStreets, courtesy of US Census Bureau, TIGER.

\Tutorials\Chapter13.gdb\PittsburghStreets\PittsburghStreets_ND, courtesy of US Census Bureau, TIGER, and W. L. Gorr.

\Tutorials\Chapter13.gdb\PittsburghStreets\PittsburghStreets_ND_Junctions, courtesy of US Census Bureau, TIGER, and W. L. Gorr.

\Tutorials\Chapter13.gdb\Garage, courtesy of W. L. Gorr.

\Tutorials\Chapter13.gdb\GraffitiCalls, courtesy of The City of Pittsburgh, Department of Innovation and Performance, downloaded from the Western Pennsylvania Data Center, with Creative Commons license at https://data.wprdc.org/dataset/311-data. W. L. Gorr assigned random point locations within correct neighborhoods to the neighborhood-level data, added attributes (GraffitiTypeCode, RemovalTypeCode, Artist, Comment, and several dates used for scheduling graffiti removal in fictitious scenarios), and entered fictitious values.

\Tutorials\Chapter13.gdb\GraffitiCallsForCollector, courtesy of The City of Pittsburgh, Department of Innovation and Performance, downloaded from the Western Pennsylvania Data Center.

\Tutorials\Chapter13.gdb\Heading, courtesy of W. L. Gorr, Carnegie Mellon University.

\Tutorials\Chapter13.gdb\Pittsburgh, courtesy of The City of Pittsburgh, Department of City Planning.